P9-CML-290

Green and Clean Energy

What You Can Do

GREEN ISSUES IN FOCUS

Sherri Mabry Gordon

Enslow Publishers, Inc.
40 Industrial Road
Box 398
Berkeley Heights, NJ 07922
USA

http://www.enslow.com

Library of Congress Cataloging-in-Publication Data

Gordon, Sherri Mabry.
    Green and clean energy : what you can do / Sherri Mabry Gordon.
       p. cm. — (Green issues in focus)
    Includes bibliographical references and index.
    Summary: "Read about the history of energy use, renewable and non-renewable energy,
global warming, "green" houses and communities, and what you can do"—Provided by
publisher.
    ISBN 978-0-7660-3348-1
    1. Renewable energy sources—Juvenile literature. 2. Green technology—Juvenile literature.
3. Clean energy industries—Juvenile literatrue. 4. Sustainable architecture—Juvenile literature.
I. Title.
    TJ808.G67 2011
    333.79—dc22   2010000828

Printed in the United States of America
052010 Lake Book Manufacturing, Inc., Melrose Park, IL

10 9 8 7 6 5 4 3 2 1

To Our Readers: We have done our best to make sure all Internet addresses in this book were active
and appropriate when we went to press. However, the author and the publisher have no control over
and assume no liability for the material available on those Internet sites or on other Web sites they
may link to. Any comments or suggestions can be sent by e-mail to comments@enslow.com or to
the address on the back cover. Every effort has been made to locate all copyright holders of material
used in this book. If any errors or omissions have occurred, corrections will be made in future
editions.

♻ Enslow Publishers, Inc., is committed to printing our books on recycled paper. The paper in
every book contains 10% to 30% post-consumer waste (PCW). The cover board on the outside
of each book contains 100% PCW. Our goal is to do our part to help young people and the
environment too!

Illustration Credits: Associated Press, pp. 44, 92; © Jeremy Edwards/iStockphoto.com, pp. 97, 119;
© iStockphoto.com, #489380, p. 75; © jane/iStockphoto.com, p. 64; © Joseph Luoman/
iStockphoto.com, p. 13; National Aeronautics and Space Administration, pp. 23, 114; © Nancy
Nehring/iStockphoto.com, p. 101; © Kyu Oh/iStockphoto.com, p. 29; © ParkerDeen/iStockphoto.
com, p. 34; Photos.com, pp. 78, 105, 111; Shutterstock.com, pp. 5, 56, 67, 71, 83, 115, 117, 118; White
House photo, p. 19; © Lonnie Springer/iStockphoto.com, p. 51; © Ivars Linards Zolnerovichs/
iStockphoto.com, pp. 10, 113.

Cover Illustration: Shutterstock.com.

Contents

The Energy Buzz

"Turn off the lights!"

"Shut the door!"

"Turn off the TV!"

"Close the refrigerator door!"

"Shut off the computer when you are done!"

"Don't let the water run!"

Most kids across the country have heard a grown-up say something like this before. But parents, teachers, and other adults are not just nagging kids. They actually are trying to conserve energy. Standing and staring blankly into the refrigerator or leaving the front door wide open during the winter wastes energy.

Most kids are used to hearing their parents tell them to shut the fridge and turn off the lights.

There are several problems with wasting energy, says Christy Radanof, a substitute teacher in Pickerington, Ohio, and mother of two. Aside from the fact that it wastes money that could be spent on other things, energy waste causes us to use up our energy supply faster than we should and it pollutes our environment, she says.[1]

Radanof, who also owns a company that sells environmentally friendly cleaning products, says she wants young people to understand that changing simple things can have a big impact. For this reason, Radanof has not only developed a green program at the elementary school where she works but she also has been teaching her own children how to conserve energy.[2]

"They're kind of like my green police now, blowing the whistle on me every time I forget something," she says. "So, for example, they will walk by us and remind us if we leave the water on for any length of time. Whether we are rinsing dishes or brushing our teeth, the water is off until we need it—not running the whole time."[3]

But conserving water is not the family's only focus. The Radanofs are always looking for opportunities to conserve energy. For example, they bring their own bags to the grocery, combine their errands to save gasoline, and replace their lightbulbs with environmentally friendly versions.

They also recycle as much as they can. In fact, recycling has become a family activity. Radanof says they have even made it into a game to see if they really need to put the garbage can at the curb each week. Sometimes they are able to go more than two weeks without putting the garbage can out because they reuse or recycle the majority of their waste.[4]

"The little choice of 'do I throw it in the trash can or do I just take it to the recycle bin,' is training your brain," Radanof says. "If we can all do these little things and get our brains trained to do the small things then our eyes will open up to other things we can do to conserve energy. As a result, it will seem like a natural progression of what we should be doing."

"And if everyone can do just a few little things, I am a firm believer that it will have a huge impact on the health of the planet," she adds.[5]

Why Is Energy Use Important?

Energy plays an important role in everyday life. From lighting our homes to fueling our cars and powering our factories, energy is used to make life convenient and efficient—and more fun. Energy is used when we go to the movies, put batteries in our toys, and play arcade games. Energy also helps us improve ways to care for people and help them live longer.

Without energy, our world would not look the same. For instance, our homes would be dark and cold. And we would not have all the gadgets that make life easier. No one disagrees that energy is an important part of the lifestyle people are used to. But wasting energy can impact our environment at a quicker rate than necessary.

Every year Americans consume more energy than any other country in the world, with 40 percent of that energy coming from oil.[6] Our energy supply is dwindling because most of the energy our country depends on comes from nonrenewable resources like coal and oil.

When a resource is called nonrenewable, that means that it cannot be replaced. Once we use it up, it is gone forever. Nonrenewable resources like oil and coal are also called fossil fuels. Fossil fuels are made from fossilized remains of things that died years ago.

For instance, coal is formed from giant plants that died in swampy areas many years ago. Over the years, the plants are buried under water and dirt. The heat and pressure from these layers helps turn these dead plants into coal. Today, coal is the most plentiful fossil fuel produced in the United States. And the United States has the world's largest known coal reserves. But if we keep using coal at the same rate, it will be gone in about 225 years.[7]

About 92 percent of coal is used to create electricity. Coal is also used to make plastics, fertilizers, and medications. People who mine for coal use giant machines to remove coal from the ground. There are three methods used to get to the coal: surface mining, mountaintop removal mining, and underground mining, which is sometimes called deep mining.

Surface mining is used when the coal is less than two hundred feet underground. Mountaintop removal mining scrapes off the top of a mountain until the coal is visible. Underground mining is used when the coal is buried deep underground; some of these mines are a thousand feet deep.

Mining coal can have a significant impact on the environment. It can destroy land and pollute water near the mine. For this reason, restoring land damaged by surface mining has become part of the mining process. When mining is complete, dirt and rock should be returned to the area and replanted. At that point, the land can be used for another purpose. The coal industry is also working to develop clean coal technologies. These technologies remove coal's sulfur and nitrogen oxide components, which are important air pollutants.

Because the burning of coal for electricity production is responsible for almost half of all global warming gases released in the atmosphere, experts are urging that we use cleaner forms of energy production.

About 58 percent of oil products used in the United States comes from other countries.[8] Oil, which is sometimes called petroleum, is formed from the remains of plants and animals, covered with layers and layers of mud. The heat and pressure from these layers turns the remains into oil.

Oil is usually found underground in reservoirs. Scientists are able to find oil by studying rock samples and by taking measurements. Some of the tools they use include satellites, sensors,

Train cars carrying coal. Most of the energy used in the United

and global positioning devices. These tools allow them to find oil more easily. As a result, when looking for oil, they drill fewer holes than they did in the past.

Drilling for oil can disturb land and ocean habitats and can cause oil spills. Areas disturbed by drilling are sometimes called footprints. Some steps taken to reduce the impact of drilling include using moveable drilling rigs and smaller "slim hole" drilling rigs. As a result, oil drilling footprints are one-quarter the size of those thirty years ago.[9]

Additionally, with the "rigs to reefs" program, some drilling rigs in the ocean are toppled over and left on the seafloor. They then become artificial reefs that attract fish and other sea life. In six months to one year, the toppled reef is covered with all types of sea creatures including clams, coral, and sponges.

How Did We Get Here?

The Industrial Revolution, which occurred between the 1700s and late 1800s, was an exciting time. People were creating machines and equipment that made life easier and more efficient. For instance, steam engines, batteries, and electric-generating power plants were all developed in this time period. American inventor Thomas Edison invented the incandescent lightbulb and many other electrical devices during this time too. And when an internal combustion engine, which used gasoline, was developed, the age of the automobile was born. During World War I, oil was used to power ships.

All these things—electricity, cars, and manufactured goods—allowed people to focus their talents in other areas. For instance, people no longer had to weave their own cloth or make their own clothes. They could go to a store and purchase what they needed. And with cars and trains, they also could get places quicker than ever before.

But all these things require energy, the majority of it from oil and coal. Burning these fossil fuels adds carbon dioxide and other emissions into the atmosphere. Sometimes these emissions are called greenhouse gases, because they trap heat in the atmosphere (like heat trapped by glass in a greenhouse). Experts say this is making the earth get warmer.

A Lot of Hot Air

Not all greenhouse gases are bad. In fact, gases that occur naturally help regulate the earth's temperature. This natural greenhouse effect allows energy from the sun to be radiated as heat. This helps warm the planet to a temperature where life can survive.

Many scientists say that burning fossil fuels is increasing Earth's natural greenhouse effect. Heat cannot escape into the outer atmosphere, and the earth's temperature rises.

They think that deforestation may be contributing too. Deforestation involves cutting down and burning trees. Burning trees releases carbon dioxide into the atmosphere. What's more, cutting down trees means there are fewer trees to absorb carbon dioxide from the atmosphere.

The National Aeronautics and Space Administration (NASA) says that the earth's average temperature has increased by 1.2 to 1.4 degrees Fahrenheit in the last hundred years.[10] What's more, eight of the earth's warmest years have occurred since 1998. The warmest year was 2005. And if the emission of greenhouse gases continues to increase, scientists predict that

In the Amazon, millions of trees have been cut down to make room for buildings and farms. Deforestation is one of the factors contributing to global warming.

the average temperature could increase from 3.2 to 7.2 degrees above 1990 levels by the end of this century.[11] This warming of the earth is often called global warming.

Global Warming and Climate Change— Understanding the Difference

People often think that global warming and climate change mean the same thing. But they do not. Global warming is an average increase in the temperature of the earth's atmosphere. Both natural and human activities can contribute to global warming. However, when experts talk about global warming, they usually mean warming caused by human activities. Running factories, constructing houses and buildings, and driving cars, trucks, and buses are some of the human activities that can contribute to warming.

Meanwhile, climate change means more than a significant change in temperature. It also takes into account changes in wind and precipitation. According to the National Academy of Sciences, "climate change" is becoming the preferred term because it communicates that there are other changes taking place besides increasing temperatures.[12]

Changes in climate can occur for a variety of reasons. For instance, the climate may change because

*   there are changes in the sun's intensity or changes in the earth's orbit around the sun

✳ there are changes in the natural processes within the climate system, such as changes in ocean circulation

✳ there are human activities that change the atmosphere's makeup, such as burning fossil fuels, deforestation, and urbanization.[13]

Recently, the U.S. government developed a policy to address climate change. Its goals are to slow emissions, strengthen science and technology, and to get countries around the world to cooperate more in addressing climate change.

A Changing World

It is no secret that climate change affects people, plants, and animals. For instance, allergies have been linked to changes in climate for years. And diseases like malaria may be sensitive to the climate too. Recently, scientists have observed other changes as well. These include a rise in sea level, shrinking glaciers, and earlier blooming trees.

Did You Know?

DID YOU KNOW that more than half the energy-related emissions come from power plants and one third come from transportation?[14]

Growing Allergies

Many people point out that increased carbon dioxide in the atmosphere may enhance plant growth. While this trend might be good for farming, recent research shows that it can worsen respiratory allergies because it can also increase pollen in the air. Additionally, scientists have noticed that trees tend to bloom earlier as the climate changes.

"We're starting to have what we call 'season creep,'" says Brenda Ekwurzel, a climate scientist with the Union of Concerned Scientists. "Trees are blooming earlier in the spring, and the frost is coming later in the fall, extending the time for allergy sufferers to experience symptoms."[15]

"There's no denying there's a change," adds Paul Ratner, an immunologist with American College of Allergies. "It's definitely bad news for people who have allergies. It doesn't help that warming will also ... worsen asthma"[16]

In fact, the World Health Organization estimates that 300 million people in the world have asthma, and 250,000 die from the disease each year. That rate has increased a lot over the past few decades.[17]

Additionally, researchers have shown that "elevated levels of carbon dioxide stimulate weeds to produce pollen out of proportion with their growth rates ... and the weediest species seem to thrive ... in high levels of carbon dioxide."[18]

Although scientists are still unable to predict exactly how climate change will affect allergies, recent data suggests that warming will make things worse.[19]

For instance, at the end of summer in 2005 the amount of ice in the Arctic "was the smallest seen in 27 years of satellite imaging, and probably the smallest in 100 years." Some experts conclude that this is the strongest evidence of global warming in the Arctic so far.[20] What's more, the Greenland ice sheet is also melting at a much faster rate than it was before. In fact, the melting has increased 30 percent in thirty years. It is now losing more ice each year than it gains in new snow.[21]

Finally, another report says that at least seventy species of frogs are now extinct due to climate change. Camille Parmesan, a biologist at the University of Texas who conducted the study, adds that one hundred to two hundred animals that depend on the cold are in deep trouble, including penguins and polar bears. Cold-dependent species on mountaintops have nowhere to go, she adds.

"We are ... seeing species going extinct," she says. "Now we've got the evidence, it's here. It's real. This is not just biologists' intuition. It's what's happening."[22]

According to Douglas Futuyma, a professor of ecology at State University of New York in Stony Brook, biologists thought the harmful effects of global warming were further down the road. But they are finding that may not be the case.

"I feel as though we are staring crisis in the face," Futuyma says. "It's not just down the road somewhere. It is ... hurtling toward us. Anyone who is 10 years old right now is going to be facing a very different ... world by the time that they are 50 or 60."[23]

2

Things Are
Heating Up

When former Vice President Al Gore lost his bid for the presidency in 2000, many wondered what he would do next. Several years later, there was no question. When he released *An Inconvenient Truth*, a documentary about global warming, it was clear he was determined to warn others that the planet needed help or it would change forever.[1]

This documentary, which starred Gore, was a big success. In fact, it was one of the top money-generating documentaries in history, taking in more than $46 million. And the book he wrote to go along with the film was also a hit. It sold millions of copies and even made it to the top of the *New York Times* bestseller list. What's more, Gore won a Nobel Peace Prize and an Academy Award for his film.

Former U.S. Vice President Al Gore has led the fight against global warming.

But some say that perhaps the film's biggest accomplishment was raising awareness of climate change. Until Gore released his film, there was very little attention focused on the issue. Less than one month after the film hit theaters in Australia, a Lowry Institute poll found that global warming was now a major concern. In fact, most Australians saw it as a bigger priority than terrorism. Moreover, the poll found that a large portion of Australians wanted something done about climate change—even if it affected the economy.[2]

Inconvenient Facts or Fiction?

In the midst of the film's success though, some scientists have expressed concerns about the documentary's facts. They argue that some of the main points contain errors or make exaggerated claims. They also worry that Gore may have "gone beyond scientific evidence."[3]

Their main concern is not about whether or not the earth is getting warmer, but what will happen if it does. For instance, Dr. Kevin Vranes from the Center for Science and Technology Policy Research at the University of Colorado says that while he is glad Gore "got the message out," he is concerned that he "oversells our certainty about … the future."[4]

Gore talks of spikes in temperatures, melting ice caps, rising seas, and dying people. He also points to Hurricane Katrina as one example that storms are going to get larger and more destructive. "Unless we act boldly," he warns, "our world will undergo a string of terrible catastrophes."[5]

"We need to be careful in describing the hurricane story," says Dr. James E. Hansen, an advisor to Gore and director of NASA's Goddard Institute for Space Studies.[6]

Another discrepancy that experts point out is the rise in sea level Gore mentions. For instance, a report by the United Nations Intergovernmental Panel on Climate Change estimates that the world's seas will rise a maximum of twenty-three *inches* in this century—down from earlier estimates. Meanwhile, Gore says the seas will rise up to twenty feet. He also portrays New York, Florida, and other areas sinking beneath the waves.[7] "Climate change is a real and serious problem.... [But] the screaming ... does not help," says Bjorn Lomborg, a scientist from Denmark. He added that the United Nations panel does not want to scare people. What is needed he says is "careful analysis and sound policy."[8]

Meanwhile, there are special interest groups that do not agree. For instance, Cooler Heads Coalition developed a Web site designed to "dispel the myths of global warming." Their Web site states that while global warming is real and carbon dioxide emissions are contributing to it, it is not a crisis.

"Global warming in the 21st century is likely to be modest, and the net impacts may well be beneficial in some places," they say. "Even in the worst case, humanity will be much better off in 2100 than it is today."[9]

The coalition notes that death rates related to extreme weather have declined by more than 98 percent since 1920. As far as the future goes, the coalition says that rising sea levels in the 21st century will be measured in inches rather than feet.[10]

Disagreements over global warming and climate change do not end with Gore and his documentary. Scientists have been debating the specifics for years. They disagree on the level of impact and what we should be doing about it.

A Heated Debate

There are some scientists who say that there is not enough evidence to even support the idea of global warming. They maintain that the increase in temperatures is part of a normal cycle within the earth's atmosphere and that fluctuations are nothing to be concerned about.

For instance, Dr. S. Fred Singer, an atmospheric physicist at George Mason University and founder of the Science and Environmental Policy Project, says that global warming scenarios are alarmist. He also maintains that "computer models reflect real gaps in climate knowledge and future warming will be inconsequential or modest at most."

"Climate change is a natural phenomenon," Dr. Singer says. "[Our] climate keeps changing all the time. The fact that [our] climate changes is not in itself a threat, because, obviously, in the past human beings have adapted to all kinds of climate changes."

He also says there are problems with climate models used to predict the effects of global warming. "You have to also understand there's something like two dozen climate models in the world," Dr. Singer says. "And one question to ask is: 'Do they agree?' And the answer is: They do not."

"These models are all produced by excellent meteorologists, fantastic computers. [Then], why do they not agree?" he asks. "Well, there's a reason for this. These models differ in the way they depict clouds.... In some models, clouds produce an additional warming. In some models, clouds produce cooling.

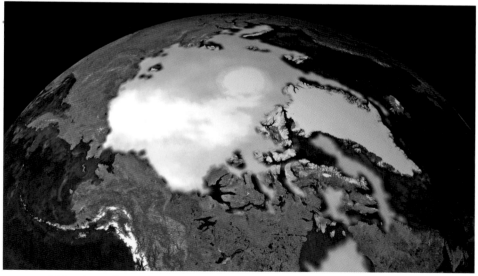

This photo from the National Aeronautics and Space Administration, or NASA, shows the shrinking polar ice cap.

Which models are correct? There's no way of telling. Each modeler thinks that his model is the best."[11]

Meanwhile, Fred Palmer, president of Western Fuels Association, Inc., an industry lobbying group, believes that "many scientists, politicians and environmental groups have greatly overstated the threat and consequences of climate change." He says:

> In the ten years I've been involved in it, there is no basis, no mechanism that anybody can point to or look at to say that more carbon dioxide in the air is going to lead to catastrophic global warming ... as opposed to some mild warming, which is nothing to be concerned about at all.... Reasonable people are concerned, but to me the concerns are speculation and not based on observations or on any scientific mechanism that they can point to.

Palmer also argues that coal is an important part of the United States' economic success. If the country were to cut back on its use of fossil fuels, he says, it "would seriously affect the world's social and economic progress." He maintains:

> In the past, we've had ... struggles over pollution in the United States.... [But] carbon dioxide is a benign gas required for life on earth. It is not a pollutant. There are no state laws dealing with carbon dioxide.... So when [environmentalists] say, "We live too well, or there are too many of us, we have to cut back in this area or that area, we have to put less greenhouse gases in the atmosphere," they come first to [the coal industry], because we are not only the

biggest source of carbon dioxide, we're the biggest source of electricity. And … the low-cost electricity that the coal plants provide has enabled our society to have the economic success that we have."[12]

(It should be noted that Palmer is the representative of an industry lobbying group, not a scientist.)

There are also a few groups that disagree with the majority of scientists. They believe global warming could actually be beneficial instead of harmful. Two examples include the Greening Earth Society in the United States and the Subtropical Russia Movement in Russia. For instance, the Greening Earth Society maintains that increased carbon dioxide is good for plants. The society believes that more carbon dioxide will produce better, faster-growing plants. They also do not think that carbon dioxide should be considered a harmful emission.

Richard C. J. Sommerville, a climate modeler with the Scripps Institution of Oceanography who has written extensively about climate change, says the American people should be concerned about climate change.

"Concerned" is a good word for climate change. Not alarmed, and not [indifferent]. So far as we know, this is a phenomenon with a long time scale. On the other hand, we have to keep in mind that there have been surprises in the past. The ozone hole is a wonderful example. There was a theory that ozone would be slowly depleted if our use of ozone-depleting chemicals such as air conditioning refrigerants was not drastically reduced. Indeed, the discovery that half the ozone over the Antarctic atmosphere disappeared every southern spring was a huge surprise. The Montreal protocol resulted in a shift towards more environmentally benign chemicals, and the hope is that the ozone hole will restore itself in the next few decades. And there's a lot of recent evidence that the climate system is capable of behaving like a switch rather than a dial, and producing surprises.[13]

Power Struggle

The concern over climate change has resulted in a global focus on finding new ways to provide energy. Resources like the sun, wind, water, and waste products are being investigated because they are renewable. If an energy resource can be replaced, that means it is renewable. Unlike oil and coal, renewable energy will not run out and it is better for the environment because it does not cause pollution or damage the earth. For this reason, renewable energy is also called "clean energy" or "green power."

The Environmental Protection Agency (EPA) says green power is energy "that provides the highest environmental benefit." In other words, electricity is considered green when it can be produced with no human-caused greenhouse gas emissions. Additionally, the EPA does not consider nuclear power a source of

The Kyoto Protocol

Despite the disagreements over global warming, the scientific consensus opinion has resulted in nations recognizing that something must be done to address energy use and emissions. As a result, an agreement was reached by the United Nations Conference on Climate Change in Kyoto, Japan, in 1997.

The agreement is known as the Kyoto Protocol, and it went into effect February 16, 2005. A total of 141 countries accepted and signed the treaty—but not the United States, Australia, or Monaco. Through the agreement major industrial nations pledged to reduce their emissions of greenhouse gases between 2008 and 2012. Even though China and India, the world's largest polluters, signed the treaty, they do not have to address emissions until after 2012 because they are considered developing countries.

The American delegation signed the protocol in 1997, but the Senate has refused to ratify the treaty. They feel the targeted reductions are so strict that it would have a big economic impact on the country.[14]

green power because it requires mining and long-term storage of radioactive waste.[15]

Solar Power

Perhaps one of the most powerful forms of renewable energy is solar power, or power from the sun. In fact, just one hour of sunlight has enough power to meet world energy demand for a year. [17] Solar power can be used to provide heat and it can be used to make electricity. When it is used as a heat source, water and air in buildings and homes are heated by using solar panels on the roof.

Solar power can also be used to make electricity. A solar-powered calculator is a common example of using power from the sun to operate something. But getting energy from the sun is expensive and very inefficient (we can only convert 10 percent of incoming solar radiation into useful energy such as electricity). Experts are still trying to find cost-effective ways to use energy from the sun.

Wind Power

Using the wind to make electricity is not a new idea. Windmills, which are often found on farms, are examples that have been around for a long time. But this type of windmill can make only a small amount of electricity. As a result, power companies are building what is called a wind farm to harvest wind. A wind farm

Solar panels capture energy from the sun.

has lots of huge wind turbines. These turbines are built on flat, open areas where the wind blows at least fourteen miles per hour or offshore in the ocean where wind blows almost constantly.

When the wind turns the blades of a windmill, it spins a turbine inside a generator to produce electricity. Currently, wind farms produce enough electricity to meet the needs of more than six hundred thousand families in the United States. Hawaii has the largest wind turbine in the world. It is twenty stories tall and its blades are as long as a football field.[18]

Biomass Power

The word *biomass* means "natural material" and includes materials like wood, paper, sludge, and other organic waste products. Places

Did You Know?

Did you know that anything that does work uses energy? For example, playing soccer, growing a tomato plant and listening to your MP3 player all require energy. What's more, energy comes in many different forms and can be turned into different forms and stored.

that contain a lot of this biomass, such as farms and landfills, have become significant producers of biomass power. Nearly half of the renewable energy in the United States comes from biomass. In fact, biomass power provides enough electricity for 2 million homes.[19]

Burning biomass was the most common way to capture energy. But research has shown that now there are more efficient and cleaner ways to use biomass. For instance, it can be converted into liquid fuel by cooking it using a process called gasification to produce combustible gases.

In Iowa, one company recycles more than 150 tons of biomass material every day to create electricity for 4,000 homes. The biomass they use includes leftover wood, cornstalks and corncobs, and paper and cardboard that cannot be recycled in other ways. Meanwhile, in Wisconsin people are using their local trash dump to create energy. As the trash breaks down or decomposes, it gives off a gas called methane. A machine then captures the methane and uses it to produce electricity. A similar process can be used to turn the methane gas from animal waste or cow manure into electricity. The United States Environmental Protection Agency has helped convert 360 landfills into energy producers. The organic materials in the landfill are broken down by microorganisms into methane, which is converted into electricity using generators.

Hydropower

Around the world, hydropower or waterpower is the most commonly used renewable energy resource. In fact, hydropower provides enough power for 28.3 million people. And it has been around for awhile.

The first hydroelectric power plant opened in the United States in 1882. And by the 1940s nearly half of the country's electricity came from hydropower. But after World War II, coal power plants became more popular.

Today, hydropower provides about 10 percent of the electricity in the United States. Most of the country's hydroelectric power plants are in California, Oregon, and Washington. The Grand Coulee Dam in Washington is the biggest hydroelectric dam in the United States. Construction on this dam began in 1933 and was not finished until 1942. Today, it is the largest concrete structure ever built. It is 5,223 feet (1,595 meters) long and 550 feet (167.6 meters) high.[20]

Geothermal Power

Geothermal power is found under the earth. Volcanoes and geysers are examples of geothermal power. The lava and steam that come from volcanoes and geysers come from underground heat. To use geothermal power, pipes are buried more than four feet underground. These pipes carry liquid that absorbs the heat and brings it back to the surface to heat buildings. In the summer, the system works in reverse by absorbing heat in the building or home and moving it back into the earth. Geothermal heating is very energy efficient because almost no energy is wasted.

Geothermal power can also be used to make electricity. A power plant taps into the steam or hot water underground to produce electricity. The world's largest geothermal power plant is the Geysers power plant in California. It produces electricity for more than twenty-two thousand homes.[21]

Addressing the problems in our environment involves using renewable energy. But that is not the only answer. We also need to conserve the nonrenewable energy we use by building and using more energy-efficient houses, communities, products, and cars.

Put a Cap on It

Some experts believe that one way to encourage people to conserve energy and address climate change is to charge businesses for carbon emissions. Businesses may be more motivated to use renewable energy when they are charged for polluting the environment.[22]

One way to do this is through a cap-and-trade system. This system puts a cap or ceiling on how much carbon a company can produce. Meanwhile, companies that produce less carbon than allowed can sell credits to those who are over the limit.

The Chicago Climate Exchange (CCX) was developed based on this idea. The CCX, which is the world's first carbon market, functions like the stock market, except that emission rights are traded instead of stock.

Participating companies are required to reduce emissions by 6 percent between 2000 and 2010. If a company reduces emissions more than that, they can sell their extra credits on the market. In turn, companies that cannot make the necessary reductions can buy someone else's excess.

"Let's say you're particularly good at reducing emissions.... And let's say I can't. I buy your reductions until I'm capable of doing it," explains Richard Sandor, founder of CCX. "[The country gets] the same system wide reduction, but it's done by those who can do it most cost-effectively and quickest."[23]

The price of carbon started out at a dollar a ton. But during the 2008 election year the price went from $1.90 to $7 because all of the candidates favored a national cap-and-trade system.[24]

The belief behind the cap-and-trade system is that carbon levels will fall. Greener companies can make money. And people will invest in renewable energy.[25]

Waste Not, Want Not

There is no denying that the world's easily available supply of oil and coal is decreasing and renewable energy provides promise to reduce our dependence on fossil fuels and reduce global warming emissions at the same time. But most experts agree that the best way to address the world's dwindling supply of fossil fuels is to become more efficient with what we use.

For instance, the U.S. Environmental Protection Agency says that the average home spends about nineteen hundred dollars on energy every year. But if every household in the United States replaced just one lightbulb with a compact fluorescent light, they would save money and reduce emissions. In fact, the pollution that would be prevented would be equal to removing one million cars from the road.[26]

Experts say that America has wasted energy for years.[27] However, now that prices are going up, people are looking for ways to save energy—and save money. They have found that renewable energy not only can save them money, but can also help the environment.

According to McKinsey Global Institute (MGI), we could cut the world's energy demand in half by 2020 by cutting waste in simple ways such as turning off lights in office buildings at night. MGI also says that if companies spend the $170 billion to make improvements like green buildings and cars that get better mileage, that in return could create $900 billion per year in savings by 2020.[28]

Finally, experts agree that conserving energy will buy us the time we need to perfect renewable energy sources and make them more affordable. Ultimately, the long-term goal is that green energy will replace fossil fuels as our primary energy source.

"The biggest source of immediately available 'new' energy is the energy that we waste every day," says Samuel Bodman, U.S. Department of Energy secretary.[29]

3
. . . .

Home, Green Home

Nestled in the quiet community of Washington Court House, Ohio, Craig and Amanda Pickerill built their dream house—a 3,900-square-foot (362-square-meter) structure that is the first "green" home in their area. Pickerill says they had three goals in building their new home. They wanted it to be energy efficient, to have clean air, and to be built with renewable materials.

"For instance, all the wood we trimmed with was cut down thirty years ago. So we didn't cut down any trees to [build our home]," says Pickerill. "I found a lady whose father had cut the wood out of Hocking Hills in the '70s and he just had it stored in his barn since then. We then went and had it milled at a local mill."[1]

Pickerill says they also wanted to avoid using materials that give off harmful gases. Many products give off gas when they are new, he says. This released gas is actually chemicals used in making the materials. (This process is called "off-gassing.")

"We have no carpet in our house and we have no vinyl in our home," he explains. "Vinyl is soft and it off-gases for years. It's just like when you get a new car. That new car smell is off-gassing of all the chemicals."[2]

To achieve energy efficiency, Pickerill says they built an Energy Star home. Energy Star house certification is based on an energy review of the home during the building process, he says. A company reviews the home's systems and building materials to determine the rating.

"We received Energy Star's Five Star Plus rating, which is the highest rating possible," he says. "Our house is about twice as energy efficient as the typical new-build house."[3]

For instance, the Pickerills estimate that they save about $150 per month on their heating bills. And their gas bill is only fifteen dollars per month. The savings they are experiencing are slightly more than U.S. Department of Energy (DOE) estimates. The DOE says that building energy-efficient homes has saved homeowners an average of four hundred dollars per year. Meanwhile, the extra construction costs have been as little as five hundred dollars more to build an energy-efficient home.[4]

Craig Pickerill believes that a great deal of their savings is due to the type of systems they chose for their home. For instance, their water heater is extremely efficient because it is an on-demand system.

"It only heats water when you need it," he explains. "I think the water heater is about 30 percent of the electric bill [in traditional homes]. So it saves a lot of money."[5]

The Pickerills also have a dual-source heat pump, which helps save money too. A heat pump transfers heat from the outside air to warm the home, rather than creating it using fuel. During milder winters, it costs less to heat a house with a heat pump than with a furnace that burns fossil fuels.

Energy Star

Energy Star is a symbol used for energy efficiency. For this logo to appear on products, they must meet standards set by the U.S. Department of Energy and the Environmental Protection Agency.

The goal of the Energy Star program, which was developed in 1992, is to reduce both pollution and energy bills. Energy Star products include refrigerators, dishwashers, clothes washers, lights, and so on. For instance, Energy Star washers use 50 percent less energy per load than regular washers. An Energy Star washer can save more than 9,440 gallons of water per year in every household. That is more water than the average person drinks in a lifetime.

Finally, if just one of every ten households purchased Energy Star appliances, the change would be the same as planting 1.7 million new trees.[6]

Using a heat pump is efficient until the outside temperature gets too cold. Then the systems become inefficient, Pickerill says. At that point, it allows the gas furnace to take over.

"I think at around 20 degrees, heat pumps begin to cost more to run," he says. "That's when our heater switches over to gas heat. So dual-source means that we use a heat pump at warmer temperatures and gas heat when it becomes colder."[7]

To achieve the goal of clean air and increase energy efficiency, Pickerill says they concentrated on "sealing the envelope" of their home.

"When you seal the envelope of your home, you want to have such good insulation that your house can't really breathe," he says. "Then, you use a heat exchanger, which brings outside air in and exchanges it with inside air."[8]

Heat exchangers also help with the heating and cooling process, he says. For instance, the exchanger takes the warmth out of the air you are blowing out and puts it into the cooler air that is coming in.

The overall structure of the Pickerills' home also makes it greener and helps them use less energy. For instance, their home is better insulated than traditional homes—even in the basement.

> We used the Colorado Green Building Checklist to build our house. I chose it because it is a point system that made it really simple. For instance you get more points if you choose a high level

insulation system. So we built our home using structural insulated panels. Those are called SIPs and it is a really easy way to build an energy-efficient home.[9]

SIPs have an outside wall made of concrete. Meanwhile, the middle of the wall is continuous foam insulation. Pickerill compares the assembly of the walls to that of "putting a kid's toy together." The challenge, he says, is finding a builder who is comfortable with the construction.[10]

"It is hard to find people to build your home. They are not comfortable building with these new systems," he says. "They have been building houses out of wood, or stick and frame homes, for their whole careers. When you tell them you want to build out of these other things, they think you are crazy."[11]

Their basement is unlike those found in traditional homes too. Rather than having it made out of poured concrete or block, their basement was made in a factory with insulation built into it. It took a crane to install it.

"Concrete is a really poor insulator," Pickerill says. "You lose a lot of heat through your basement."[12]

Finally, the Pickerills have long-term plans to use renewable energy to heat their home.

"We put a standing seam metal roof on our house and conduit all the way to the attic from the basement. The rear of our house faces south too," he says. "Our plan is to add solar panels in the next year."[13]

Green Building Movement

Building a new home is exciting. From fresh paint and new flooring to a new design and location, people enjoy seeing a home come together. However, building a new home can be hard on the environment. For instance, new home construction creates nearly 7 million tons of waste every year. In addition, 38 percent of the country's carbon emissions are produced by buildings—both commercial and residential.[14] For this reason, more and more people are looking into ways to build greener homes.

"Green homes are definitely on the radar screens of consumers," says Christine Ervin, CEO of the Green Building Council. "Within ... years, many more people will be aware of the comfort and affordability of green homes."[15]

The main idea behind the green home building movement is to use resources efficiently. This means not only using renewable energy like solar, wind, and geothermal power but also choosing materials that do little, if any, harm to the environment. An example might include using reclaimed or salvaged lumber rather than fresh lumber to build walls and support beams. There are some companies that specialize in getting building materials from older homes that are about to be torn down. Instead of discarding the materials, this lumber can be put back into new construction.

Other ways to make sure resources are used efficiently and the new home construction is as green as possible include:

* making sure the home does not use a lot of energy—this can be done with everything from using efficient appliances like Energy Star to incorporating renewable energy like solar power into the design

* using recycled and responsibly harvested materials with fewer chemicals (for instance, a bamboo floor is much easier on the environment than a vinyl floor)

* creating cleaner, more breathable air—aside from using equipment to ensure the air quality is top-notch, certain materials like paint and flooring also impact air quality

* making sure the impact on land and water is as low as possible—for instance there are lots of ways to minimize the use of water in the home including everything from collecting rainwater to reducing the amount of water used and controlling the energy used to heat water.[16]

Building greener homes is catching on. Despite a lagging housing market, there has been a 500 percent increase in the number of green homes in the last six years. What's more, green home building accounts for nearly $2 billion annually in the United States. It is expected to grow by at least $6.5 billion by 2010.[17]

Home builders say that increasing energy costs, consumer demand for green homes, and the better performance of green products are driving the green building movement. In fact, 82

Zero Energy Homes

A Zero Energy Home (ZEH) combines energy-efficient construction and appliances with renewable energy systems, such as solar power. The result is a home that produces its own energy and sometimes more energy than it needs. Even though the home might be connected to a utility grid, it uses "net zero energy."[18]

One benefit of a ZEH is that the home's construction reduces temperature fluctuations. It also increases reliability because it can be designed to work even during blackouts. Finally, a ZEH saves energy and reduces pollution.

The U.S. Department of Energy has partnered with building professionals and organizations to further develop the ZEH concept. To date, only a few of these homes have been constructed and researched.

A workman holds a solar panel to be installed in a zero energy home—one that generates as much energy as it uses.

percent of builders say energy efficiency tops the list of green methods requested.[19]

And the potential for the green building movement is huge. In fact, it is expected that about 1.5 million new homes will be built and furnished each year.[20] Just how many of those new homes will be built "green" is uncertain. But experts are optimistic.

"All homes will one day be green," predicts David Johnston, author of *Building Green in a Black and White World*. "It makes too much sense for it not to happen."[21]

Green Power and the Home

Experts expect to see more people, like the Pickerills, making their dream homes green. They also anticipate that in the coming years, small-scale renewable energy projects such as installing solar panels on the roof or purchasing small wind turbines are going to become increasingly popular.

But for people who cannot afford to build a new home or remodel an existing one, there are other options. For instance, some homeowners are able to buy green power directly from an electricity provider. In fact, more than 50 percent of people in the United States now have an option of purchasing a green power product.[22]

Many states also have implemented electricity competition, which makes purchasing green power even easier. Under

Setting a Green Example

The Smart Home featured at the Chicago Museum of Science and Industry is an example of a green home. Featuring everything from a recycled plastic deck to bathroom tiles made out of recycled wine bottles, the 2,500-square-foot (232-square-meter) home is an example of environmentally responsible living. The home, which shows visitors how easy it is to be green, also includes a solar paneled roof and 300-gallon barrels to collect rainwater.

"We tried to look for ideas in every choice that we make in our homes ... hoping that everyone who goes through it will be inspired to make some change on some level," says Michelle Kaufmann, the architect who designed the Smart Home.[23]

Museum officials say the goal of the exhibit is to show people that everyone can save energy and conserve resources— "whether it is an entire house or a single feature."[24]

"One thing that is fundamental to green building is that it can look like anything," says David Johnston, who owns an international green building consulting firm in Colorado. "It can be a regular Craftsman house or a Cape Cod house ... or an adobe house in Santa Fe. You don't have to change what the home looks like to make it green."[25]

electricity competition, consumers can choose who they buy their electricity from. As a result, they have the freedom to choose a provider that emphasizes a commitment to renewable energy.

According to the DOE, "by choosing to purchase a green power product, you can support increased development of renewable energy sources."[26] This support in turn then reduces the burning of fossil fuels, such as coal and oil. The DOE Web site lists organizations that offer green power.

Finally, before taking on a huge remodeling project or hoisting a wind turbine in the yard, experts recommend people look first at conserving energy and using energy more efficiently. Energy conservation involves changing behavior, like air-drying clothes instead of using a clothes dryer. Energy efficiency involves using less energy to do the same thing, like installing a more energy-efficient lightbulb to light a room. Therefore, if you want to help preserve the environment and save money, it makes good sense to do all of the reasonable conservation and efficiency measures first.

After all, most people use more energy at home than anyplace else. Energy is used to heat and cool the house, to heat water for cleaning and bathing, and to power electronic devices such as televisions, DVD players, and computers.

Energy Vampires

Some electronics use power even when they are turned off. These items are called "energy vampires." For instance, a cell phone charger that is left plugged in is still sucking in small amounts of energy even though it is not charging anything. Other vampires include televisions, computers, and printers. Even your electric toothbrush is an energy vampire. It is using energy just by being plugged in.

Although the energy cell phone chargers and electric toothbrushes are using may seem small, when all the wasted energy is added up, it equals about $4 billion a year in the United States alone. And the Department of Energy says that about 75 percent of the energy used to power home electronics is used when the products are turned off.[27]

To rid your home of energy vampires, consider unplugging the following items:

* handheld vacuums

* VCRs

* DVD players

* automatic coffeemakers

* televisions

* empty appliances

* power drills and electric screwdrivers

* computers and printers

Another option is to purchase a power strip and plug everything into the strip. Then all you have to do is flip a switch and the power is cut off. Craig Pickerill says another option is a smart outlet.

"It completely turns your television off when it's not on," he says. "There are lots of things people can do to save energy. You don't have to build a new house to cut down on your energy use."[28]

Did You Know?

DID YOU KNOW that across the United States, home refrigerators use the electricity of twenty-five large power plants every year?[29]

Experts agree that getting rid of energy vampires is one of the easiest ways to conserve energy and protect the environment. Plus, it can save up to 5 percent a month on your electric bill.[30]

Carbon Footprints—Are You Bigfoot?

Your carbon footprint is determined by adding up the amount of carbon dioxide produced from all your activities. Transportation, time spent using electronics, where you eat, and what you eat are all considered. For example, when you use a carbon footprint calculator you would answer questions about how you get to school, where you eat most of your meals, and how you spend your time.

Carbon footprints are used to measure one's impact on the environment. Having a small carbon footprint is better than being Bigfoot. Unfortunately, most Americans have big feet when it comes to carbon footprints. In fact, on average, each person in the United States is responsible for about twenty-two tons of carbon dioxide emissions every year, according to statistics compiled by the United Nations. That is far bigger than the world average of six tons.[31]

The best way to reduce the size of your carbon footprint is to use less energy.

What About Carbon Offsets?

The idea behind carbon offsets is that you reduce the amount of carbon dioxide you put into the environment by "offsetting" the carbon emissions you cannot reduce on your own. What this means is that once you have conserved energy and cut back as much as you can, you then pay a group like CarbonFund or TerraPass to help reduce the size of your carbon footprint. These companies then invest in projects like developing renewable energy, improving energy efficiency, and implementing tree-planting projects.

Each year, Christy Radanof and her family contribute to an organization that plants trees across the country. "[We do this to] offset our carbon emissions for our vehicles. It costs us about $35 a year to take care of both of our cars."[32]

However, there is some disagreement over the accuracy of carbon offsets. For instance, some experts argue that projects taken on by companies offering carbon offsets do not really compensate for the carbon the person is releasing into the air.[33]

"It is very unproductive to leave people with the impression that we could possibly plant our way out of the problem," says Joe Romm, an expert on carbon offsets.[34]

From Lightbulbs to Light Trucks

The United States is realizing how important it is to conserve energy and become more energy efficient. As a result, in December 2007, President George W. Bush signed a law that requires automakers to develop more fuel-efficient cars. It also calls for improved energy efficiency in refrigerators, freezers, dishwashers, and even in lightbulbs. In fact, the law states that lightbulbs will need to become 70 percent more efficient.

Replacing regular lightbulbs with compact fluorescent bulbs is a good way to save energy.

"I firmly believe this country needs to have comprehensive energy strategy," said President Bush.[35]

The new law focuses on energy conservation on everything from "lightbulbs to light trucks," said Representative John Dingell of Michigan. And the new lighting standards alone will lower annual electricity bills by $13 billion in 2020, remove the need for sixty mid-size power plants, and reduce emissions by 100 million tons a year, says the Alliance to Save Energy.[36]

In 2009, the corporate average fuel economy (CAFE) standard set by the Obama administration for model year 2011 is 27.5 mpg for passenger cars and 24 mpg for light trucks.[37]

Green Living

Because so much of our world revolves around energy, we are confronted with lifestyle decisions that affect our environment every day. And the choices we make can impact the quality of life for those who follow us. We use energy in almost every aspect of our lives. It is used in making and transporting goods. And it is used when we get rid of things we no longer need.

One of the biggest choices that any of us will make is how we will live each day. For instance, will we live any way we want, not considering the impact it has on the world? Or, will we consider our purchases and how we dispose of our waste?

Almost everyone is familiar with the "three Rs"—reduce, reuse, and recycle. In fact, the phrase has been used so much that it is now commonplace. But even though the message has been around awhile, it still applies. The three Rs are the key to saving energy and living a greener lifestyle. They include:

* **Reduce waste.** This is done by making smart decisions when purchasing products, including trying to buy products without a lot of extra packaging. It also involves buying things that have been made from recycled products.

* **Reuse materials.** This involves looking for creative ways to reuse containers and products rather than pitching them in the trash can.

* **Recycle.** Lots of things can be recycled—everything from paper and plastic to food scraps, yard trimmings, and electronics. Recycling can also involve selling or giving away used clothing and furniture instead of dumping them in the garbage.

The world is becoming more aware of the need for energy conservation and protecting the environment. As a result, it is getting easier to incorporate the three Rs into everyday life. For

Did You Know?

DID YOU KNOW that every time you open the refrigerator door, up to 30 percent of the cold air can escape?[38]

instance, products made from recycled material are becoming increasingly popular and are easier to find in the marketplace.

What's more, in 2006, the United States recycled 32 percent of its waste, according to the Environmental Protection Agency. This is the energy equivalent to saving more than 10 billion gallons of gasoline.[39]

The Benefits of the Three Rs

* ✳ Save at least 2,400 pounds (1,089 kilograms) of carbon dioxide a year by recycling at least half of your household waste.

* ✳ Save 1,200 pounds (544 kilograms) of carbon dioxide by avoiding heavily packaged products and cutting down your garbage by 10 percent.

* ✳ Reduce your use of fertilizers and water by creating homemade compost and using it on your yard.[40]

Eating Green

Most of us probably do not think about where our food came from when we sit down to eat. But where and how our food is grown can make a difference on the environment. Energy is used to grow, transport, package, and cook our food. What's more, some production methods pump more carbon dioxide

THE EASY ENERGY ACTION PLAN

10 SIMPLE WAYS TO USE ENERGY WISELY

1	Turn off lights.	☐ CHECK THE BOX
2	Use energy-saving light bulbs.	☐
3	Shut off computers.	☐
4	Use "smart" power strips.	☐
5	Turn off entertainment devices when not in use (TV, game systems, etc.) **OFF**	☐
6	Use natural light, heat and cooling.	☐
7	Unplug your phone charger when not in use.	☐
8	Talk to your parents about ENERGY STAR® appliances.	☐
9	Talk to your parents about programmable digital thermostats.	☐
10	Talk to your parents about Home improvements (windows, doors, roofs, etc.)	☐

U.S. DEPARTMENT OF **ENERGY**   LoseYourExcuse.gov

Shopping for local produce, and carrying it in a reusable cloth bag, helps our eating habits have a lower impact on the environment.

into the environment than others. So in order to "eat green," a lot of factors must be considered.

For instance, eating locally produced foods is typically easier on the environment. But not always, warn some experts.

"Food miles are a great indicator of localness," but not of environmental impact, says Rich Pirog, the associate director of the Leopold Center for Sustainable Agriculture at Iowa State University.[41]

In other words, local food can be best if it is purchased "in season." In other words, purchasing a locally grown watermelon in central Michigan in the middle of winter, when it is not in season, may actually require more energy than purchasing it someplace else. Growing the watermelon locally might take more energy than it takes to transport it from a warmer climate. In the middle of winter, energy would be needed to heat, light, and irrigate a greenhouse.

"You can't just look at the transportation piece," says Gail Feenstra, a food analyst at the University of California. "It's one piece of the whole puzzle."[42]

How food is grown and harvested also plays a role. For instance, some experts say that "New York state apples can be less eco-friendly than those imported from New Zealand, where growing conditions produce greater yields with less energy."[43]

According to the organization Living Green, there are a number of benefits of eating green:

✱ The average American dinner travels 1,500 miles (2,414 kilometers) before reaching the dinner plate. Eating

Did You Know?

DID YOU KNOW companies that make new products from recycled materials use 30 percent less energy?[44]

Beware of Greenwashing!

More and more people are "going green." Consequently, some companies are trying to cash in on this movement by using a tactic known as "greenwashing." Greenwashing is created from the term whitewashing. When someone whitewashes something, he is trying to make it seem more innocent than it really is. With greenwashing, a company spends more time and money claiming to be green than actually becoming green.

An example of greenwashing is the hotel chain that claims to be green because it encourages guests to reuse towels and sheets rather than having them changed every day. But that is all the chain is doing. While reusing towels does save water, there is much more the company could be doing. In other areas of their business, they are doing very little to save water and energy.

In 2007, TerraChoice, a science-based marketing firm, focused attention on greenwashing with a study known as The Six Sins of Greenwashing. Their study found that 99 percent of the more than one thousand products they reviewed were guilty of greenwashing.[45] They categorized the various forms of greenwashing into "six sins," which include:

1. Sin of the Hidden Trade-Off: An example would be "energy-efficient" electronics that contain hazardous materials. Fifty-seven percent of all environmental claims committed this sin.

2. Sin of No Proof: An example would be shampoos claiming to be "certified organic" but with no verifiable certification. Twenty-six percent of environmental claims committed this sin.

3. Sin of Vagueness: Examples include products claiming to be 100 percent natural when many naturally occurring substances are hazardous, like arsenic and formaldehyde. This sin was seen in 11 percent of environmental claims.

4. Sin of Irrelevance: Examples include products claiming to be CFC-free, even though CFCs were banned twenty years ago. This sin was seen in 4 percent of environmental claims.

5. Sin of Fibbing: Examples include products falsely claiming to be certified by an internationally recognized environmental standard like EcoLogo, Energy Star, or Green Seal. This sin was found in less than one percent of environmental claims.

6. Sin of Lesser of Two Evils: An example would be organic cigarettes or "environmentally friendly" pesticides. This sin occurred in one percent of environmental claims.[46]

The president of TerraChoice, Scott McDougall, says that the goal of the study was to help shoppers find the truth. Most of the time, the claims made about products were not actual lies, but they were very misleading to the average consumer.[47]

For instance, some garbage bags that are sold are labeled "compostable." What this statement implies is that a person can throw the bags into a compost bin or a landfill and the bags will break down quickly. However, plastic bags like these still take years to break down—unless they are sent to an institutional composting facility. Once at the facility, the bags must be stirred with huge tractors to create heat and to speed the breakdown process. But most consumers do not know this is what is meant by "compostable." They read the label and believe that the bags will turn into dirt without any intervention.

"Consumers are inundated with products [like this] that make green claims," McDougall says. "Some are accurate, certified and verifiable, while others are just plain fibbing to sell products."[48]

local food [can reduce] the consumption of fossil fuels and wasteful packing materials.

✳ Buying locally keeps money in your community.

✳ Local farmers' markets, which sell items that are grown in season, reduce the amount of energy required to grow and transport the food to you by one-fifth.

✳ [Buying organic can be beneficial] because soils capture and store carbon dioxide at much higher levels than soils from conventional farms.

✳ If we grew all of our corn and soybeans organically, we would remove 580 billion pounds (263 billion kilograms) of carbon dioxide from the atmosphere.[49]

"It's no longer a question if people want to go green or not. They do," concludes Michelle Kaufmann. "People are wanting an alternative."[50]

GO GREEN!

What you can do to help

Your parents do not have to build a new home, nor do they have to take on a large remodeling project, in order to conserve energy. There are a number of things that can be done around the house that will save energy—and money. Aside from shutting off lights, getting rid of energy vampires, lowering the thermostat, and replacing lightbulbs, here are some other ideas:

✳ **Ask your parents if you can put a plastic bottle filled with water or pennies in the toilet tank.** The weight of the water bottle will cause the tank to use less water every time you flush. In fact, *Plenty* magazine estimates that it will save about a quart to half a gallon per flush.[51]

✳ **Make sure the lint filter on your clothes dryer is clean.** The dryer becomes more energy efficient and runs better without lint clogging it.

✳ **Reuse batteries when you can.** You can stretch the life of your batteries if you switch them before you pitch them. For instance, batteries that no longer power a flashlight might still work in a television remote.

✳ **Turn off dripping faucets every time you see them.** One drop per second can add up to 165 gallons (625 liters) of water a month—that is more water than one person uses in two weeks.[52]

4
. . . .

Street Smart

The Kirk family from Wylie, Texas, does not own a hybrid vehicle—at least, not yet. But they are doing as much as they can to conserve energy and pump less carbon dioxide into in the environment. For instance, they own just one car instead of two cars like most American families. They also carpool, combine errands into one trip, and encourage their daughter, Katie, to bike to school.

"Sometimes it is inconvenient trying to conserve gasoline," admits Jennifer Kirk, who moved from Ohio to Texas several years ago with her husband, Matt, and daughter, Katie. "But we try hard not to go driving around unless we have multiple things to do. For example, I'll delay grocery shopping until I have to go to the bank across the street too."[1]

Kirk, who is a prekindergarten teacher and a full-time college student, is creative when it comes to conserving gasoline. Aside from trying to schedule her college courses to correspond with her carpool buddy, she also looks for ways to combine errands with her friends and family.

"When there were sales for teachers at the office supply stores, a bunch of us got together to go all at once," says Kirk. "[My husband] Matt and I also do our errands together instead of separately."[2]

And although the Kirks do not own a hybrid car, they investigated purchasing one when they purchased their Ford Escape.

Kirk said:

> We looked at hybrids when we bought our last car in 2006, but for the size we need they are too expensive. From the research we did, you save on gas. But if they break down all the parts are very expensive. So it didn't seem to be a true savings over time. The compact cars, like Prius, are more reasonably priced. They are great if you don't travel seventeen hours to Ohio twice a year [to visit family]! We only own one car. So it is important our family fit in it with all our stuff.[3]

A Gallup Poll conducted in 2006 supports the Kirks' research. In fact, the report states that although "hybrid cars offer breaks ... the math doesn't always add up to ... savings."[4] Take

Riding a bike to school saves gas.

the Ford Escape as an example. The nonhybrid version like the Kirks own is much cheaper than the hybrid version of the Escape. And according to GreenerCars.org, it is not that hard on the environment either. In fact, the nonhybrid version received a green score of thirty-eight in comparison to the forty-two of the hybrid version.[5]

Yet despite all the questions regarding savings, 57 percent of Americans say they would consider buying a hybrid when replacing the vehicle they now own. What's more, younger people find hybrids more attractive than older Americans. Sixty-four percent of people eighteen to forty-nine would consider purchasing a hybrid.[6]

Not All Hybrids Are Created Equal

When people think about conserving energy and protecting the environment, they often think of hybrid vehicles first. Hybrid cars have a battery-powered motor that works with a gasoline-powered engine. In top hybrids like the Toyota Prius, owners drastically reduce the amount of gasoline they use. But not all hybrid vehicles are created equal. In fact, only half of all hybrid models are more fuel efficient than nonhybrid versions of the vehicle.[7]

According the Union of Concerned Scientists (UCS), these hybrids are "hollow hybrids." These vehicles, like Saturn's 2007 hybrid, Chevy's 2007 Silverado, and GMC's 2007 Sierra pickup truck, do not have the technology or the efficiency to be called a hybrid.

The UCS points to "muscle hybrids" as problems too. These hybrids use "the battery to boost the power of a big engine rather than increase fuel efficiency," says the UCS.[8] The cars that fall into this category include the Lexus line of hybrids and

Did You Know?

DID YOU KNOW that the oil we burn through transportation releases nearly 2 million tons of carbon dioxide into the atmosphere each year?[9]

the Honda Accord hybrids. Honda has since ditched the Accord hybrids. According to spokesperson Chris Naughton, Honda has learned a lot and is now focusing on fuel efficiency.[10]

Plugging in to Fuel Efficiency

Some experts believe that emission-free electric cars are the wave of the future. But so far the batteries that operate the cars cannot supply the energy needed.

However, General Motors may have found a compromise with their new Chevy Volt. The Volt has an electric motor and a battery that allows drivers to travel forty miles (sixty-four kilometers) on one charge. If the driver goes over forty miles, then a gas engine kicks in to recharge the battery while driving. As a result, the driver can go several hundred more miles before needing to fill up or recharge.

"With [past electrics], people had to change the way they lived," says Andrew Farah, the Volt's chief engineer. "I want a vehicle that doesn't ask them to change at all."[11]

Because 80 percent of people drive less than forty miles a day, people driving the Volt would use little if any gasoline. The company predicts that Volt will cost two cents per mile to drive while using the battery, saving the average person fifteen hundred dollars annually. What's more, they say the Volt will be "less expensive to recharge than purchasing your favorite coffee."[12] Sales on this sedan are set to begin in 2010. And there

The Chevy Volt is slated to go on sale in 2010.

is an unofficial waiting list of more than thirty thousand people eager to buy them.[13]

But some experts are concerned that the cars will not be fully tested by then and even if they are, the price could be too high. If the Volt costs as much as forty thousand dollars, as some experts predict, drivers will find it hard to earn that back in gas savings, they say.[14]

Wind Beneath Our Wheels?

Meanwhile, other experts believe air-powered cars may become more attractive to consumers than electric cars. While both cars do a great job of cutting emissions, what makes a compressed air vehicle (CAV) more attractive than electric cars is the price. The CAV is expected to sell for about twenty thousand dollars, while electric cars can cost twice as much. CAVs work by compressing air instead of gasoline to move engine pistons.

The CAV also has a number of fuel options. For short commutes and in town driving at up to thirty-five miles per hour, the car can go sixty miles on a tank of compressed air. For freeway driving, the CAV can travel at legal speeds for eight hundred miles using a small motor that compresses outside air to keep the tank filled. Meanwhile, the motor can burn just about anything—gasoline, ethanol, even cooking oil.[15]

Experts predict that these cars may be built starting in 2011. The factories will make about eight thousand cars a year and sell them directly to consumers. The only question yet to be answered is: How will these cars do in crash tests? Many CAVs are small and made of light material that may not hold up as well in a crash as steel does.

Pain at the Pump

Aside from the fact that consumers want to go easy on the environment, the price of gasoline is also contributing to the popularity of hybrids, electric cars, and CAVs. In fact, in the last ten years, the price of gasoline has quadrupled. As a result, Americans are driving fewer miles than ever, so less gas is needed to power our cars.

In fact, "U.S. gasoline demand will likely decline in 2008 for the first time in more than seventeen years," says a report by Cambridge Energy Research Associates. "For the first time since the 1970s and early 1980s the number of miles driven by Americans has clearly begun trending downward."[16]

The report also says that pickup trucks, minivans, and sport utility vehicles (SUVs) are less popular than ever before. In fact, sales for these vehicles have fallen below 50 percent for the first time since 2001. Higher gas prices are blamed for this decrease as well.[17]

More than 75 percent of Americans are convinced that the rise in the price of gas they have experienced is permanent. And more than half of Americans believe that the price of gas will reach six dollars a gallon over the next five years. To deal with

Did You Know?

DID YOU KNOW that the Federal Highway Administration estimates that it costs people between twenty-two and twenty-nine cents per mile to drive a car, depending on its size? But by carpooling every day, people can save up to three thousand dollars a year on gas, insurance, parking, and wear and tear on their car.[18]

the rising prices, 71 percent say they are considering buying a more fuel-efficient vehicle.[19]

"With climate change concerns now, it's very likely that fuel efficiency will be at the forefront for the foreseeable future," says Samantha Gross, one of the authors of the Cambridge report. "It's unlikely we will go back to not caring about fuel efficiency the way we did in the late 1980s."[20]

Is Homegrown Fuel the Answer?

Many experts feel that biofuels, fuels made from plant materials, are part of the answer to addressing rising fuel costs, global warming, and our country's dependence on foreign oil. In fact, experts believe that our dependence on oil from other countries is also impacting our nation's security and its economy. Many think that biofuels can help free Americans from this dependence. Biofuels also create rural jobs and improve opportunities for farmers. And if the technology is perfected it could also impact global warming.

Currently, ethanol, which is made from corn kernels, is the primary biofuel being used in this country. But because producing ethanol competes with corn for food products, can lead to deforestation, and requires lots of water, experts are looking for other types of biofuels. For instance, prairie grasses such as switchgrass, which occur naturally and are not food products, can produce a lot of cellulose. Cellulose can in turn be made into biofuel. However, the challenge is that cellulose is a lot more difficult to break down and turn into ethanol than corn. Other options include municipal trash, agricultural waste, algae, and even carbon dioxide. But none of the technologies have been demonstrated to be feasible.

As a result, some say biofuels may be doing the opposite of what was hoped. For instance, using land to grow fuel causes the destruction of forests, wetlands, and grasslands. What's

Higher prices at the pump are causing many people to think about getting more fuel-efficient cars, such as hybrids.

more, the trees and plants in these natural areas store enormous amounts of carbon. But when they are destroyed, the amount of carbon released into the atmosphere increases.

"It turns out that the carbon lost when wilderness is [cut down] overwhelms the gains from cleaner-burning fuels," writes Michael Grunwald, in "The Clean Energy Scam," an article in the April 7, 2008, issue of *TIME* magazine.[21]

Also, the large demand for farm-grown fuels has raised world food prices and has endangered the hungry. For instance, Grunwald says, "the grain it takes to fill an SUV with ethanol could feed a person for a year."[22] As a result, the U.N.'s World Food Program says the rising cost of food is a global emergency and they need $500 million in additional funding and supplies. And Lester Brown of the Earth Policy Institute says that "biofuels pit the 800 million people with cars against the 800 million people with hunger problems."[23]

Also, a study in *Science* magazine found that when deforestation is considered, corn ethanol and soy biodiesel produce twice the emissions of gasoline. Even cellulosic ethanol increases overall emissions when its plant source is grown on good land.[24]

"The lesson behind the math is that on a warming planet, land is ... incredibly precious ... and every acre used to [make] fuel is an acre that can't be used to [make] the food needed to feed us or the carbon storage needed to save us," Grunwald writes.[25]

Addressing Fuel Efficiency

The government has also started to take fuel efficiency more seriously. In December 2007, President George W. Bush signed legislation requiring automakers to address fuel efficiency.

"We make a major step … toward reducing our dependence on oil, fighting global climate change, expanding the production of renewable fuels and giving future generations … a nation that is stronger, cleaner, and more secure," President Bush said.[26]

Under the new law, automakers will have to increase fuel efficiency by 40 percent. This increase will require cars, SUVs, and small trucks to get thirty-five miles (fifty-six kilometers) per gallon by 2020.

Experts said the fuel efficiency requirements will save people seven hundred to a thousand dollars a year in fuel costs. It will also reduce oil demand by 1.1 million barrels a day, they say.[27]

Some auto industry experts believe that more than half of vehicles will need some type of hybrid technology by 2020 to meet these new standards. But they say that more reasonable alternatives will be needed. Many customers are not willing to pay the higher price for a hybrid. Hybrids can cost as much as four thousand to five thousand dollars more than a nonhybrid car.[28]

The law also calls for an increase in ethanol use to 36 billion gallons (136 billion liters) a year by 2022 with at least 21 billion gallons (80 billion liters) coming from something other than corn. When the law was signed, only 6 billion gallons (23 billion liters) of ethanol were being used in a year, and it was primarily made from corn. Former President Bush predicted that this requirement will cause companies to develop ethanol from prairie grass and wood chips.

Overall, the law is intended to cut U.S. oil demand by 4 million barrels a day by 2030. This reduction is more than twice the country's daily imports from the Persian Gulf, experts say.[29]

GO GREEN!

What you can do to help

You may be wondering: "What can I possibly do? I don't even own a car!" But you do not have to own a car or even be able to drive to conserve energy. You do not even have to convince your parents to buy a hybrid to have a positive impact. In fact, the small steps you take toward energy conservation can go a long way in reducing emissions. Here are some ideas:

* **Skip the drive-through window and go inside.** According to Environmental Defense, when a car idles for more than ten seconds it actually uses more gas and creates more pollution than restarting the engine.[30]

Did You Know?

DID YOU KNOW that idling at a drive-through window and in stop-and-go traffic costs motorists 753 million gallons (2,850 million liters) of gasoline a year, or $1,194 per driver in wasted fuel and time?[31]

One way to cut down on pollution and gas consumption is to go into a restaurant rather than using the drive-through.

✳ **Carpool with neighbors or friends or ride the bus.** For instance, the National Safety Council says that one full forty-foot bus takes fifty-eight cars off the road.[32]

✳ **Encourage family and friends to drive slower.** Not only are slower speeds safer, but also the EPA estimates that every five miles per hour you drive over sixty miles per hour costs another twenty cents per gallon of gas.[33]

✳ **Ask your parents to remove the car's roof rack.** The Rocky Mountain Institute estimates that you can save fifteen to thirty gallons of gasoline a year by just leaving the rack off half the time.[34]

✳ **Get your mom or dad a gift certificate for a tune-up on their next birthday.** According to the EPA, a well-tuned engine can save thirteen cents a gallon. A clean air filter can save thirty-two cents a gallon, and properly inflated tires can save ten cents a gallon.[35]

5

. . . .

Knowledge Is Power

Teaching young children about the importance of caring for the earth is at the core of Christy Radanof's Go for the Green Challenge program. Radanof, who developed the program for Tussing Elementary in Pickerington, Ohio, believes that if children learn to care for the world around them it will become an automatic part of their life—as natural as tying their shoes or brushing their teeth. She says she also wants children to feel empowered. Even though they are young, kids can still make changes that will have an impact, Radanof says.

"As a parent, I saw Earth Day come and go several years in a row with very little if any mention in the classroom. And as a substitute teacher, I understand that there is so much information that teachers have to get through and only so much time

WE
RECYCLE

Students celebrate Earth D
Many believe that learn
to take care of the earth
help young people trea
more responsibly.

in the day to do it," Radanof explains. "I wanted to develop a program that would be simple to incorporate into their day, wouldn't require more than a couple minutes a day, but that would encourage and empower the kids with the knowledge that they can create change."[1]

The Go for the Green Challenge contains a number of different elements. For instance, each classroom selects a "green ambassador" to be the liaison between the classroom and Radanof. Additionally, the green ambassadors keep their classmates on task and develop ideas on what they could do to impact the environment. They also make sure the classroom recycles paper and reduces its energy use by turning off lights and equipment.

Meanwhile, teachers were presented with a number of different tools. Radanof developed worksheets and arranged for the kids to become members of the EPA Kids Club. Each teacher received a list of books for their classrooms and Web sites the classes could visit. Radanof also arranged for the teachers to use the materials provided by MeetTheGreens.

"*Meet The Greens* is a cartoon ... about different energy and environmental issues," Radanof says. "The teachers could show the cartoon and then there would be a quiz afterwards that the kids could take as a group and learn a little more."[2]

The goal for each classroom was to meet all the challenges on their list and win a Green Tiger Paw for their classroom.

"Everybody was kind of competing with each other to see who would get the Green Tiger Paw first," explains Radanof. "In the end, every single classroom ended up with the Green Tiger Paw. That was really exciting because I had set a goal that 60 percent of the school would get their Green Tiger Paw. But everyone ended up on board!"[3]

As a reward, each student got a tree to plant at home. Initially, Radanof had planned just to give each classroom a couple of trees and draw for the winners. But she was so pleased with the results that she wanted everyone to be rewarded.

"I wanted each kid to see immediate benefits for doing something for the environment," she explains. "Many times when you take action to help the environment, you don't see results for awhile. I wanted the kids to see some immediate benefit for their efforts."[4]

Green Inspiration

Radanof says her inspiration for the challenge came from Dr. Wangari Maathai, an environmentalist and activist. In 2004 Maathai became the first African woman to receive the Nobel Peace Prize for "her contribution to sustainable development, democracy, and peace," says Radanof. She explained:

> When she was growing up, Dr. Maathai noticed that the water, crops, and animals were disappearing in her small South African country. And she

had this brainstorm that there was a connection between that and all the trees her countrymen were cutting down. So what she decided to do was to teach women how to plant trees. She encouraged those women to take it to the different tribes and to plant trees. As a result, the landscape of her country was totally changed. They have water, crops, a forest and the animals are back. What amazed me was that she was one person who took it upon herself to encourage others.[5]

"We can all have that same impact if we share our ideas," says Radanof. "Throughout the challenge, I make a strong point that kids need to communicate their ideas. Kids have fantastic ideas, but no one can act on them if they don't share them with others."[6]

What's Driving You?

Twenty-three million kids ride school buses each year. Yet the air inside the bus may actually be worse than the air outside. For instance, Yale researchers found that the air inside diesel school buses had five to fifteen times more toxins than outside air.[7] And another study found that levels of diesel exhaust inside a school bus can be four times higher than those in cars driving just ahead of the bus.[8]

Exposure to diesel fumes is not uncommon. The vast majority of the school buses still use diesel fuel, even though less harmful fuels are available. In fact, many school districts still use old models, including buses built in the 1980s.

Changing behaviors is one way to reduce the impact buses have on the environment. For instance, cutting down on the amount of time a school bus spends idling helps reduce harmful emissions and can save money. Typically, school buses burn about half a gallon of fuel per hour when they are idling.

Experts recommend that school officials adopt policies about idling. Some examples include:

* When loading and unloading children, the school bus should be turned off.

* Early-morning warm-up should be limited to idling no more than three to five minutes except in extreme weather.

* Buses should not be permitted to idle while waiting for students during field trips and extracurricular activities.

* The "cleanest" buses should be assigned the longest routes.[9]

Recently, the EPA passed new emission standards for buses. Under the new standards, buses built after 2007 are required to release 90 percent less soot and 95 percent less emissions.

To do this, schools can either replace the buses with alternative fuel buses or retrofit existing buses. Retrofitting a bus involves putting a particle filter on the bus that helps reduce emissions. Switching fuels also improve air quality as well.

Students wait to board a bus to school. School buses can be a serious source of pollution.

For instance, the Wissahickon Pennsylvania School District switched to low-sulfur emission fuel.

"We've already started using a low-sulfur emission fuel," says Joe Malseed, the head mechanic for the district. "What will happen is as the vehicles start using it, it will clean their engines. We're hoping it will reduce emissions from 18 to 30 percent per bus."[10]

Lights Out! Where Schools Can Save Energy

Operating a school costs a lot of money. And many school districts pay more for energy than for supplies and books. In fact, K-12 schools in the United States spend more than $6 billion a year on energy. But according to the U.S. Department of Energy, at least a quarter of that could be saved by conserving and managing energy better. This would cut the nation's school energy costs by $1.5 billion each year.[11]

One way schools can save money is by changing behavior. By simply turning off lights and computers, schools can save thousands of dollars. In fact, lighting is responsible for 50 percent of the electric bill in most schools.[12] Another way to lower lighting costs is to keep lights clean. Dirt and dust can reduce the amount of output from lights by as much as 15 percent a year.[13] Replacing the plastic covers, or diffusers, on lights also helps improve light output.

Exit signs are another area where lighting costs can be reduced. For example, replacing exit signs that use incandescent

Did You Know?

DID YOU KNOW that if you reduce the amount of time spent idling by ten minutes each day, you can keep 550 pounds (250 kilograms) of carbon dioxide out of the air every year?[14]

lights with light-emitting diodes (LED) can save money. LEDs can last twenty-five years without needing to be replaced, compared to one year for incandescent lights. What's more, LED lights pay for themselves in less than a year in lower energy costs.

Even making changes to vending machines can save money. When a vending machine operates continuously, it can cost a school $200 to $350 a year. However, some companies make energy-control devices for vending machines that can save schools as much as 47 percent. Turning off the lights in the vending machine is another way to save money. For instance, Seattle School District saved $20,000 a year by turning off the lights in its 250 vending machines.[16]

Another way schools can save money is through daylighting, in which sunlight is captured on the rooftop and redirected to interior spaces, often using reflective tubes. According to the Sustainable Buildings Industry Council, the average middle school that uses daylighting will save tens of thousands of dollars annually. The research also suggests an improvement in student performance. For example, one study found that students who attend daylit schools for two or more years scored 14 percent better on tests than students in schools that did not use daylighting.[17]

Finally, experts recommend that when school districts are constructing new buildings that they build green—especially since energy-efficient schools do not cost more to build. Schools that are designed to save energy and reduce their environmental impact can save the district 50 percent on energy bills compared to traditional schools.[18]

High-performance green schools typically use daylighting, have renewable energy systems, and include water conservation and recycling features. The Durant Road Middle School in Raleigh, North Carolina, is such a school. For instance, they use daylighting as well as a radiant barrier on the roof that reflects the sun's heat. As a result, the amount of air conditioning required by the school is about 30 percent below that needed by a traditional school, so that the school saves tens of thousands of dollars in energy costs each year.[19]

Did You Know?

DID YOU KNOW that only 10 percent of the electricity used by an incandescent lightbulb is turned into light? The other 90 percent is wasted as heat.[20]

It All Adds Up

One report makes the following suggestions for schools to cut energy costs:

Lighting strategies

 ✳ Schools can save anywhere from 8 to 20 percent on lighting by turning off lights in unoccupied rooms.

 ✳ Cleaning lamps and light fixtures regularly can save up to 15 percent on lighting.

Computers and Office Equipment

 ✳ Energy Star monitors have a low-power sleep mode that uses only between two and ten watts.

 ✳ Energy Star copiers can save schools 40 percent compared to standard models.

Maintenance

 ✳ Proper boiler maintenance can lead to energy savings of 10 to 20 percent.

 ✳ Timers can be installed to shut off electric hot water tanks when the building is not occupied.

Kitchen/Vending

 ✳ Schools can reduce energy consumption by preheating ovens for no more than fifteen minutes before use.

 ✳ An energy control device for vending machines can save as much as 47 percent with payback of less than two years.[21]

Solar-Powered Education

Nineteen schools in South Carolina received a solar power system in 2006 for their school and an accompanying teaching program. This system was provided by Santee Cooper and the Electric Cooperatives of South Carolina Inc. and is designed to teach students about renewable energy.

"Santee Cooper has been a leader in renewable energy," says Lonnie Carter, president and chief executive officer of Santee Cooper. "This solar school program allows us to show not only children, but educators, the effectiveness of renewable energy options and the challenges associated with meeting current energy needs."[22]

The project also includes a renewable energy curriculum. The curriculum incorporates teaching, research, and hands-on opportunities. It is also supplemented by an Internet-based monitoring system that provides information about the system's performance.

In September 2001, Santee Cooper became the first electric utility in South Carolina to generate and offer green power to its customers. They use methane gas collected at landfills as well as solar power to generate renewable energy.

GO GREEN!

What you can do to help

As a student, you have a vested interest in helping your school save money. If your school spends less money on energy, it will have more money for educational programs and activities. But you do not have to convince your school board to build a new school in order to have an impact. Little things add up too. Below are some simple ways you can encourage your school's teachers and staff members to help cut energy costs.

* **Ask your teacher if you can conduct a class experiment to see where there are drafts.** Once you have determined the drafty areas, the class can make long, thin cloth bags and fill them with beans to use as insulation along windows and doors.

* **Form an "energy patrol" with your school's permission.** Write down all the things in the school that use energy, from the lights to the refrigerator in the teacher's lounge to the computers in the library. Submit ideas for how the school can lower energy costs by turning things off when they are not being used.

* **Ask your teacher if you can print your reports on both sides of the paper.** This not only saves paper, but cuts down on the energy used printing the report.

* **Look for ways your school can reduce paper and submit your ideas to the principal.** For instance, can the school newsletter be sent via e-mail? Or can your parents send electronic notes to the teacher instead of a handwritten note?

6

Sustainable Communities

The tornado that ripped Greensburg, Kansas, apart in May 2007 was the one of the worst in U.S. history, rocketing through the town at 205 miles (330 kilometers) per hour. It was more than a mile and a half wide, and it claimed the lives of eleven people.[1] All that was left of the small town were the sidewalks and underground sewers.

After the tornado, the townspeople had a decision to make. Either they could move on, or they could rebuild Greensburg and make it stronger—and greener.

About half of the fifteen hundred residents decided to stay and build. Their commitment was to rebuild the town "as a showcase of environmentally friendly living." As a result, the town investigated ways to use renewable energy.

Businesses and residents were encouraged to build energy-efficient homes.[2]

"If you are going to build a community from the ground up, it is our responsibility to think about the future," says city administrator Steve Hewitt.[3]

"I would never say the tornado happening was a good thing. I would never wish that on anybody," says Kim Alderfer, assistant city manager. "But given the opportunity, we have to do it right—to make it better."[4]

It took some work to get everyone on board, says Daniel Wallach, who formed the nonprofit group Greensburg Green-Town. He says they showed residents that going green was about more than saving polar bears. It also involves cutting waste, saving on energy costs, and building a stronger town. Those arguments made sense.

"Our church sometimes costs up to $1,000 a month to heat," says Marvin George, pastor of a Baptist church in the town. He says he hopes to rebuild the church to meet the highest energy efficiency standards.[5]

"Self-sufficiency and independence are strong values out here. Farmers and ranchers all make their living from the land, so they have an awareness and sensitivity to it that is unique," says Wallach.[6]

And although the rebuilding process is slow, it is moving forward. Recently, the 4,700-square-foot (437-square-meter) city hall building opened. Located in the heart of the town, it has become a symbol of the community's commitment to become a model for sustainable communities everywhere. The building gathers solar energy, collects rainwater for reuse on-site and makes maximum use of daylight. They are hoping this building will receive a LEED Platinum Certification.[7]

LEED, which stands for Leadership in Energy and Environmental Design, is a rating system. It was developed by the U.S. Green Building Council and sets standards for sustainable construction. LEED Platinum buildings cost about 5 percent more to construct, says Jack Rozdilsky, a University of North Texas professor who has studied Greensburg's rebuilding

A tornado that hit Greensburg, Kansas, in 2007 caused widespread devastation. The citizens decided to rebuild their city on "green" principles.

effort. But, he says the buildings typically have 30 to 50 percent cheaper energy bills.[8]

Greensburg has already captured the nation's attention and their hearts. For instance, President George W. Bush spoke at the graduation commencement for the town's eighteen high school seniors—something he has never done before. And Leonardo DiCaprio produced a thirteen-part series about the town. It is called *Eco-Town* and aired on the Discovery Channel. What's more, the National Building Museum featured a one-year exhibit on the town's rebuilding efforts.

"They are really making a wonderful opportunity out of an absolute tragedy," says Susan Piedmont, museum curator and architect. She adds that having small towns like Greensburg embrace environmentally friendly architecture will help people see how easy environmental building can be.[9]

But what impresses people most is the town's vision and optimism despite suffering a tragic loss.

"Ninety percent of the town was just gone overnight, and yet the social fabric was intact even without the buildings," says Stephen Hardy, a city planner with the architectural firm BNIM. And it is this social fabric that allows a town like Greensburg to rebound like it has.[10]

"You don't experience a storm like this and come out of it with an attitude of complacency," says Wallach.[11] "People from around the country and around the world will come [to Greensburg] to see what the future looks like," he says.[12]

The Move Toward Sustainable Communities

A healthy community is the cornerstone of a healthy society. It is the place where people connect with one another—where they live, work, play, and learn. As a result, experts say that sustainable communities are becoming more attractive because people realize that if they work together, they can produce a higher quality of life.

According to the Natural Resources Defense Council, people are recognizing that development is "gobbling up the American countryside at an alarming rate." In fact, each hour, about 365 acres of open land is turned into "strip malls, anonymous suburbs, and traffic-clogged roads."[13] As a result, some communities are choosing to develop differently. They are looking to become sustainable.

In communities that sustain themselves, businesses, households, and government make efficient use of land, energy, and other resources, allowing the area to improve life with minimal waste and environmental damage. These communities are healthy and secure, and provide people with clean air to breathe and safe water to drink, according to the Civic Renewal Movement of the Civic Practices Network (CPN).

According to the leaders of CPN: "In sustainable communities, people are engaged in building a community together. They are well-informed and actively involved in making community decisions. They make decisions for the long term that benefit future generations as well as themselves."[14]

Benefits of Sustainable Communities

How and where we build has a huge environmental impact. According to Greener Communities, buildings account for 38 percent of annual carbon dioxide emissions, 30 percent of both raw materials waste and landfill waste, and 12 percent of potable water consumption. But if communities make a commitment to become sustainable, they can reverse these trends. This requires environmentally responsible planning and building as well as the creation of energy-efficient and water-efficient buildings. One way this is accomplished is by using renewable resources such as solar and geothermal energy. This in turn reduces the emissions of carbon dioxide.

Sustainable communities also strive to conserve natural resources by choosing materials that are renewable, recycled, and durable when constructing homes and other buildings. And they reduce waste through the reuse and recycling of materials.[15]

Did You Know?

DID YOU KNOW that nearly half of the greenhouse gas emissions in the U.S. come from buildings?[16]

Living Green—Examples of U.S. Sustainable Communities

If you look across the nation, you will see varying degrees of sustainability. Some communities have been sustainable for years while others are just getting started. Listed below are three communities—each addressing sustainability, but each at different stages.

Seattle, Washington. Seattle was one of the first cities in the United States to incorporate the ideas of sustainability into community planning. According to the volunteer civic organization Sustainable Seattle, a "sustainable city" is one that "thrives without compromising the ability of future generations to meet their needs."[17]

Experts explain that an important part of Seattle's sustainable city definition is the concept of the city as a system within a system. In other words, they want neighborhoods to do well within the city. As a result, decision making is done by considering the effects on the entire system and on future generations.

Since 1990, many new city policies, plans, and programs have included sustainability. For instance, Seattle has been working with industry in the area. As a result, the city and Boeing, a company that makes airplanes, agreed to use waste heat from a new sewer trunk line to provide heat for the company. This saves on the cost of both heating equipment and fuel.

Experts say that many of Seattle's most successful sustainability projects have been city- and county-sponsored programs. These programs train volunteers so that they can teach others about more sustainable practices. Some of the programs have included "Friends of Recycling," "Master Composters," and "Master Gardeners." What's more, many small businesses are implementing more sustainable practices, including everything from eco-retailers to bicycling carpet cleaners.

Seattle, Washington, with its familiar skyline dominated by the Space Needle, is an example of a sustainable community.

Burlington, Vermont. In 2007, Burlington was ranked as one of the top metropolitan areas by *Country Home* magazine and received the "Best Green Places" award. The magazine rated cities based on air quality, mass transit use, power use, and the number of organic producers and farmers' markets. Second place went to Ithaca, New York, and Corvallis, Oregon, took third place.

"We thought to ourselves, 'If we could live anywhere in the United States, where would be the best green place to live?'" says Grant Fairchild, managing editor.[18] Burlington received especially high marks for the way its people, businesses, and government value green living. Among Burlington's green features are the following:

* A compost facility collects food scraps from restaurants, supermarkets, and food manufacturers and sells the compost to farmers, gardeners, and landscapers.

* The area has sixteen farmers' markets, five producers of organic food, and three food co-ops.

* Car pools are used by 12.3 percent of Burlington-area commuters, about 5.6 percent of the workforce walks to work, and 4.6 percent work at home.[19]

"It's certainly an honor to be called the greenest city in America," says Betsy Rosenbluth, project director for Burlington Legacy, the sustainable city initiative. "Burlington ... really understands the connection between our environment and our economy and our social health."[20]

Chicago, Illinois. In September 2008, Mayor Richard M. Daley announced a plan for Chicago to become one of the greenest cities in the nation. The plan calls for reducing greenhouse gas emissions to three-fourths of 1990 levels by 2020. To get there, the city will address energy efficiency, using clean and renewable energy sources, improving transportation, and reducing pollution.

"We can't solve the world's climate change problem in Chicago, but we can do our part," says Daley. "We have a shared responsibility to protect our planet."[21]

Daley is one of about eight hundred mayors who have agreed to adopt that goal. But Chicago was the first to identify specific pollution sources and outline how it would achieve the reductions. For instance, the city has an agreement with two coal-fired power plants to reduce emissions or shut down by 2015 and 2017. The plan also calls for increasing recycling and car pooling, promoting alternative fuels, and expanding the number of green rooftops. Also called living rooftops, green

Did You Know?

DID YOU KNOW that about 500 billion to one trillion plastic shopping bags like the ones you get at the grocery are used worldwide each year? That means about one million bags are used each minute. And more than 380 billion of these are discarded in the United States. Less than one percent of them are recycled.[22]

rooftops contain living plants. The rooftop may contain grass, shrubs, flowers, and sometimes even trees if it can support the weight.

Rebecca Stanfield, a senior energy advocate at the National Resources Defense Council in Chicago, says the city has a lot of work ahead of them.

"It's not like you can just walk away from this and say, 'We've got a plan to do this,'" Stanfield explains. "It's a [call to action for businesses], to the government, ... [and] to the advocacy community that we've got a lot of work to do but at least we've got a road map."[23]

All in all, there is growing interest in sustainable communities. But experts say until we can get legislators, like those in Chicago, Seattle, and Burlington, to enact policies that encourage their development, it is going to be a challenge to produce green communities.

GO GREEN!

What you can do to help

As a young person, you may be wondering what you can do to help your community become more sustainable. Here are a few ideas to help you get started. Remember, communities become sustainable when each community member does his or her part.

Because plants use carbon dioxide to live, planting trees is one way to fight global warming and contribute to a sustainable future.

* Plant a tree. According to AmericanForests.org, planting trees around your home can conserve energy and lower costs. In fact, planting trees in the right places can reduce air conditioning costs by 10 to 50 percent.[24]

* Bring your own bags to the grocery. Plastic bags from groceries and other stores often end up littering community streets. They also get tangled in treetops, choke animals, and clog sewer systems. And if they ever do make it to a landfill, they take years to decompose.

* Ask your parents if you can recycle rainwater. If you use a rain barrel to hold up to 6 percent of the rainwater falling on your property, you can help reduce flooding and pollution in the storm water system.[25]

7

Red, White, and Green! Green Power and the Future of America

Global warming, climate change, and the green movement are not new ideas. In fact, these ideas have been around for awhile. Scientists have been warning Americans for years what could happen if we do not change the way we use energy. But experts say it took years for people to buy into the movement because of the way in which it was presented.

"Green-minded activists failed to move the ... public not because they were wrong, but because the solutions they offered were unappealing," says Alex Nikolai Steffen, author of *Worldchanging: A User's Guide for the 21st Century.* "They rejected technology, business and prosperity in favor of ... a simpler way of life."[1]

But, Steffen says, "you don't change the world by hiding in the woods." As a result, a new green movement is taking shape.

He says it is a movement that embraces environmental concerns but offers different answers to the problems. For instance, he says that business, technology, scientific exploration, and sustainable policies "can propel the world into a bright green future."

"Americans trash the planet not because we are evil but because the ... systems we've devised leave no other choice," he explains. "Our ranch houses and high-rises, factories and farms, freeways and power plants were [developed] before we had a clue how the planet works."

"[But] tomorrow we might see vehicles that consume no fossil fuels and emit no greenhouse gases," he predicts. "Combine cars like that with smarter urban growth and we're well on our way to sustainable transportation."[2]

Generation Green

Some experts believe that the green movement will be pushed forward by young people. It is no secret that kids can be passionate about issues. From the civil rights movement in the 1960s to the "no nukes" movement in the 1970s and outrage over child labor in the 1980s, students have the ability to change history. And many believe that it is passionate students who will really impact the government regarding climate change.

For instance, Bill McKibben, an author and environmentalist, tells students that old-fashioned marches and protests will

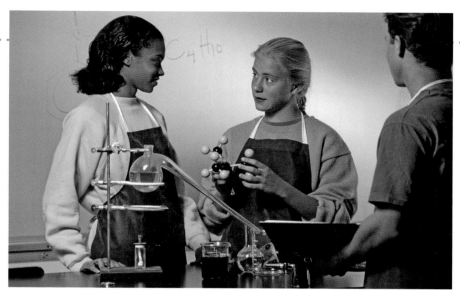

Young people are often at the forefront of societal change such as that needed to promote clean, green energy.

no longer work. Moving the government to change will require coordinated action and a smart use of the Internet.[3]

As a result, McKibben and a number of students developed a "virtual march" on Washington in April 2007. They called it *Step It Up National Day of Climate Action*, and they organized more than fourteen hundred communities to get their message out. In each location, participants held up banners that said: *Step It Up, Congress: Cut Carbon 80 percent by 2050*. Photos from these locations streamed to the Web from all across the country. Some material was even streamed from underwater in Key West, from the dwindling glaciers in Wyoming, and from the levees of New Orleans.[4]

McKibben believes that because this generation will suffer the consequences of climate change, their efforts toward a greener America will make a difference. "There are a lot of people who are educated about global warming and want to figure out what to do," he says. "The students, ultimately, are the mainstream-in-training."[5]

Green Energy Revolution

A report developed by the European Renewable Energy Council and Greenpeace says that the world could eliminate fossil fuel use by 2090 if current energy use is overhauled. This would then result in saving $18 trillion in fuel costs and create a $360 billion industry.

The report also said that renewable energy markets have almost doubled from 2006 to 2007, reaching $70 billion. As a result, the groups believe renewable energy could more than double its share of world energy to 30 percent by 2030 and reach 50 percent by 2050. Sven Teske, Greenpeace's leading author on the report says the investments would be repaid by savings in fuel costs.[6]

The report's estimates are more optimistic than those of the International Energy Agency, a group that advises rich countries. They predict only a 13 percent increase by 2030 and maintain that fossil fuels will remain the leading source of power.[7]

"[Either way] the need for energy is real—and growing especially in developing nations," Teske says.[8]

Green Price Tag

Experts disagree over how much it will cost to reduce carbon in the atmosphere. A large part of this disagreement occurs because no one can predict with certainty how bad the effects of climate change will be. Another factor is that they do not know what technologies will be developed to make reducing carbon easier. In addition, they cannot tell which policies will be implemented to make current technologies competitive.

Still, some experts have made estimates on how much going green will cost the United States. For instance, Nicholas Stern, former chief economist for World Bank, has said cutting emissions would cost one percent of the gross domestic product. He bases his number on an estimate that global temperatures will rise by 55 degrees Celsius in the next century. But other experts believe that global warming will be less disruptive and we should not invest too much money in trying to prevent something that is never going to happen.[9]

For instance, the Cooler Heads Coalition says that

> given the uncertainties, [we may] regret ... [diverting] too much of our global wealth to [solve] what may be a nonexistent or trivial problem, especially if that diversion [puts] billions in poverty. On the other

hand, we also may regret not doing anything if man-made global warming does turn out to be a problem. It is therefore prudent to examine what steps we can take that would prove beneficial whether or not ... global warming turns out to be a problem.[10]

They go on to suggest that a global warming policy is "no regrets" policy as long as it:

* Reduces greenhouse gas emissions

* Prevents or reduces global warming

* Provides methods for addressing global warming

* Does not cost a lot of money to implement[11]

Meanwhile, President Barack Obama says he wants to spend $150 billion over the next ten years on biofuels, wind, solar, hybrids, and other climate-friendly measures. The American Recovery and Reinvestment Act included more than $80 billion in clean energy investments, of which $6.3 billion was allotted for state and local renewable energy and energy efficiency efforts. Additionally, $11 billion was allotted for moving renewable energy from rural areas where it is produced to the cities where it is used most. His plan also called for 40 million smart meters to be deployed in American homes.[12]

Green Jobs

One report predicts that by 2038 this country will have produced 4.2 million green jobs. The jobs will result from the country's increased use of renewable energy and interest in energy efficiency. This increase will account for 10 percent of the new job growth.

"It could be the fastest growing segment of the United States economy over the next decade, ..." the Global Insight Inc. report says. The study was released by the U.S. Conference of Mayors.[13]

Currently, only 750,000 people work in "green jobs." These jobs include everything from scientists researching alternative fuels to makers of wind turbines and more energy-efficient products. In 2038 the report predicts that there will be 1.5 million additional jobs in finding alternative fuels for cars. Renewable power will account for 1.2 million new jobs and eighty-one thousand jobs will be found in making buildings and homes more energy efficient.

However, this job increase will not happen unless the country makes an aggressive move away from fossil fuels and toward alternative energy sources. The country also needs to tackle energy efficiency. For instance, for this prediction to become a reality, 40 percent of the electricity would have to come from renewable resources like wind and solar. Currently, renewable energy only accounts for 3 percent of the electricity.

"We [were] trying to show the size of the green jobs economy" if the country becomes less dependent on fossil fuels, says Jim Diffley of Global Insight.[14]

A Lifetime Away?

Ask any American, and you will probably find that he supports increased energy conservation. But 53 percent do not think a solution to the country's energy problems will be found in their lifetime. And nearly 80 percent believe that high energy costs are here to stay, according to a survey.

Still, consumers want the government to take more immediate steps to address the country's energy issues. Ninety-one percent of Americans feel that steps need to be taken to "dramatically increase energy conservation programs." And 65 percent say that the government should require states to teach energy conservation in driver education classes.

Americans also feel that the government should develop incentives. For instance, 52 percent feel that people who buy and drive a hybrid should get a tax break. And 44 percent feel that people who drive SUVs that are not required for their jobs should be taxed.

Finally, most Americans say they are open to changing in order to protect the environment. For instance, 61 percent said they would rather pay more for cleaner fuels than pay less for fuels that pollute. And more than 60 percent said they would accept wind terminals and solar plants in their hometowns.[15]

"The survey highlights the opportunity for the U.S. government to adopt a much more aggressive and comprehensive energy policy," says analyst Kurt Hallead.[16]

A researcher investigates the water quality of a wetlands area. The percentage of "green" jobs is expected to rise tremendously in the future.

Craig Pickerill says:

> An energy plan by the government would be much more effective at lowering energy costs—especially if people were given incentives to build houses like mine. If you could get money off your taxes for building efficiently and using renewable energy, then it would make it economically appealing for people. And if all the houses used half the energy, we would have a lot more fossil fuels available. I think it is going to be tough to completely eliminate the need for fossil fuels. But you certainly can reduce the consumption of them in the next ten years by giving people incentives to be green.[17]

Another hurdle is changing behaviors; it is always easier to continue what you are doing than to change your ways. Yet it is clear that many Americans care about the environment and conserving energy. And they seem ready to embrace green power as a means of getting there.

Chapter Notes

Chapter 1. The Energy Buzz

1. Author's interview with Christy Radanof, September 2008.
2. Ibid.
3. Ibid.
4. Ibid.
5. Ibid.
6. "Annual Energy Report," U.S. Department of Energy, July 2006, <www.eia.doe.gov/emeu/aer/pdf/pages/sec1_3.pdf> (April 5, 2010).
7. Russell McLendon, "Where Does Coal Come From," Mother Nature Network, August 4, 2009, <http://www.mnn.com/earth-matters/translating-uncle-sam/stories/where-does-coal-come-from> (December 2009).
8. "How the Climate Bill Will Reduce U.S. Imports," Environmental Defense Fund, August 26, 2009, <http://www.edf.org/article.cfm?contentID=10365> (December 2009).
9. "Climate Change—Basic Information," U.S. Environmental Protection Agency, February 16, 2010, <www.epa.gov/climatechange/basicinfo.html> (April 5, 2010).
10. Ibid.
11. Ibid.
12. Ibid.
13. Ibid.
14. Ibid.
15. Colleen Diskin, "Global Warming Lengthening Allergy Season," NorthJersey.com, April 18, 2008, <www.northjersey.com/environment/17894024.html> (April 5, 2010).
16. Bryan Walsh, "Allergies Getting Worse? Blame Global Warming," *TIME,* September 15, 2008, <www.time.com/time/health/article/0,8599,1841125,00.html> (September 2008).
17. Ibid.
18. Ibid.
19. Ibid.
20. Marc Kaufman, "Decline in Winter Arctic Ice Linked to Greenhouse Gases," *Washington Post,* September 14, 2006, <www.washingtonpost.com/wp-dyn/content/article/2006/09/13/AR2006091301817_pf.html> (September 2008).
21. "Greenland's ice sheet melts as temperatures rise," *CNN.com,* October 24, 2007, <http://www.cnn.com/2007/TECH/science/10/23/greenland.melting> (November 2009).
22. "Global warming already killing," Climate Ark, November 1, 2006, <http.www.climateark.org/shared/reader/welcome.aspx?linkid=64050> (February 2010).
23. Ibid.

Chapter 2. **Things Are Heating Up**

1. About the Film, *An Inconvenient Truth,* ClimateCrisis, n.d., <www.climatecrisis. net/aboutthefilm/> (April 5, 2010).
2. "The Gore Factor: Reviewing the impact of *An Inconvenient Truth,*" <http://www. ecosmagazine.com/?act=view_file&file_id=EC134p16.pdf> (February 2010).
3. William J. Broad, "From a Rapt Audience, a Call to Cool the Hype," *New York Times,* March 13, 2007, <http://www.nytimes.com/2007/03/13/science/13gore. html?ex=1331438400&en=2df9d6e7a5aa6ed6&ei=5090&partner=rssuserland &emc=rss> (April 2010).
4. Ibid.
5. Ibid.
6. Ibid.
7. Ibid.
8. Ibid.
9. "Global Warming 101: The Science," Cooler Heads Coalition, February 4, 2009, <http://www.globalwarming.org/2009/02/03/global-warming-101-science/> (April 5, 2010).
10. Ibid.
11. "Interview with Dr. S. Fred Singer," *What's Up with the Weather—The Debate,* PBS, 2000, <www.pbs.org/wgbh/warming/debate/> (December 2009).
12. "Interview with Fred Palmer," *What's Up with the Weather—The Debate,* PBS, 2000, <www.pbs.org/wgbh/warming/debate/> (December 2009).
13. Interview with Richard C. J. Sommerville, *What's Up with the Weather—The Debate,* PBS, 2000, <www.pbs.org/wgbh/warming/debate/> (December 2009).
14. "Kyoto and Beyond," CBC News, February 14, 2007, <www.cbc.ca/news/back-ground/kyoto/> (October 2008).
15. "Green Power Defined," EPA Green Power Partnership, U.S. Environmental Protection Agency, March 24, 2009, <www.epa.gov/greenpower/gpmarket/index.htm> (April 5, 2010).
16. "Going Green 101: Resources for Your Family," *The Oprah Winfrey Show,* April 20, 2007, <www.oprah.com/article/oprahshow/tows_past_20070420_b> (April 5, 2010).
17. "Energy Efficiency and Renewable Energy," Solar Energy Technologies Program, U.S. Department of Energy, March 19, 2009, <http://www1.eere.energy. gov/solar/animations.html> (April 5, 2010).
18. "A Few Facts About Energy," National Association of Conservation Districts, n.d., <http://www.nacdnet.org/education/resources/energy/> (April 5, 2010).
19. "Growing Energy on the Farm: Biomass and Agriculture," Union of Concerned Scientists, n.d., <http://www.ucsusa.org/clean_energy/technology_and_impacts/impacts/growing-energy-on-the-farm.html> (April 5, 2010).
20. "Grand Coulee Dam Statistics and Facts," U.S. Department of the Interior Bureau of Reclamation, January 2009, <http://www.usbr.gov/pn/grandcoulee/pubs/factsheet.pdf> (April 5, 2010).

21. "The Geysers," Calpine, n.d., <http://www.geysers.com/> (April 5, 2010).
22. Bryan Walsh, "Why Green is the New Red White & Blue," *TIME,* April 28, 2008, pp. 45–57.
23. Fareed Zakaria, "How to Blow Less Smoke," *Newsweek,* September 6, 2008, <www.newsweek.com/id/157550> (April 5, 2010).
24. Ibid.
25. Walsh.
26. Len Vermillion, "Sweeping Away Energy Waste," Manufacturing.net, May 22, 2007, <www.manufacturing.net/sweeping-away-energy-waste.aspx?menuid=242> (February 2008).
27. Walsh.
28. Ibid.
29. "DOE Seeks Applications to Invest up to $40 Million in Housing Research," U.S. Department of Energy, press release, June 13, 2007, <www.energy.gov/news/5128.htm> (October 2008).

Chapter 3. Home, Green Home

1. Author's Interview with Craig Pickerill, September 2008.
2. Ibid.
3. Ibid.
4. Len Vermillion, "Sweeping Away Energy Waste," Manufacturing.net, May 22, 2007, <www.manufacturing.net/sweeping-away-energy-waste.aspx?menuid=242> (February 2008).
5. Author's Interview with Craig Pickerill, September 2008.
6. "About Energy Star," Energy Star, U.S. Environmental Protection Agency & U.S. Department of Energy, n.d., <www.energystar.gov/index.cfm?c=about.ab_index> (April 5, 2010).
7. Author's Interview with Craig Pickerill, September 2008.
8. Ibid.
9. Ibid.
10. Ibid.
11. Ibid.
12. Ibid.
13. Ibid.
14. "Benefits for the Environment," Green Communities, n.d., <www.greencommunitiesonline.org/green/benefits/environment.asp> (April 5, 2010).
15. David J. Lipke, "Green Homes—Eco-friendly Home Building Trends," Media Central, Inc., January 2001, <http://findarticles.com/p/articles/mi_m4021/is_ISSN_0163-4089/ai_75171065> (April 5, 2010).
16. "Carbon Offsets," President Homes, n.d., <www.presidenthomes.com/Services/GreenBuilding/Default.aspx> (November 2008).
17. Ibid.

18. "Zero Energy Home Design," Energy Efficiency and Renewable Energy: U.S. Department of Energy, April 22, 2009, <http://apps1.eere.energy.gov/consumer/your_home/designing_remodeling/index.cfm/mytopic=10360> (April 5, 2010).

19. "Green Building Is Now a Trend, Maybe One Near a Tipping Point," *Mortgage News Daily,* July 10, 2006, <www.mortgagenewsdaily.com/7102006_Green_Building_Products.asp> (November 2008).

20. Lipke.

21. Ibid.

22. "Buying Green Power," *Green Power Network,* Energy Efficiency and Renewable Energy: U.S. Department of Energy, April 16, 2007, <http://apps3.eere.energy.gov/greenpower/buying/index.shtml> (April 5, 2010).

23. Associated Press, "New Exhibit has Visitors Thinking Green," CBS News, August 1, 2008, <www.cbsnews.com/stories/2008/08/01/tech/livinggreen/main4315954.shtml> (November 2008).

24. Ibid.

25. Ibid.

26. "Buying Green Power."

27. Lori Bongiorno, "What's Wasting Energy in Your Home Right Now," *Yahoo! Green,* October 9, 2008, <http://green.yahoo.com/blog/the_conscious_consumer/4/what-s-wasting-energy-in-your-home-right-now.html> (November 2008).

28. Author's Interview with Craig Pickerill, September 2008.

29. "Energy Facts," The Energy Coalition, n.d., <http://www.energycoalition.org/contents/energy-information-tips/energy-glossary.aspx> (April 5, 2010).

30. Bongiorno.

31. Clayton Sandell, "Reducing Your Carbon Footprint," ABC News, June 7, 2006, <http://abcnews.go.com/Technology/story?id=2049304&page=1> (November 2008).

32. Author's Interview with Christy Radanof, September 2008.

33. Jesse Ellison, "Save the Planet, Lose the Guilt," *Newsweek,* July 7–14, 2008, <www.newsweek.com/143701> (September 2008).

34. Ibid.

35. Associated Press, "Bush Signs Bill Boosting Car Fuel Efficiency," MSNBC, December 19, 2007, <www.msnbc.msn.com/id/22326795> (October 2008).

36. Ibid.

37. "Energy & Environment," White House Web site, 2009, <www.whitehouse.gov/issues/energy-and-environment> (April 5, 2010).

38. "Tips and Facts," Community Green: Environmental Leadership for Homeowners Associations, n.d., < http://www.caigreen.org/tips/> (April 5, 2010).

39. "Recycling," Yahoo! Green, n.d., <http://green.yahoo.com/living-green/recycling.html> (April 5, 2010).

40. "Actions for Living Green," Living Green, April 2007, <www.livinggreen.org/actions.cfm> (April 5, 2010).

41. Ellison.

42. Ibid.

43. Ibid.

44. "Did You Know?" Live Green, n.d., <http://newportlivegreen.com/LG-Did-You-Know.html> (April 5, 2010).

45. "The Six Sins of Greenwashing—New Study Finds Misleading Green Claims in 99 Percent of Products Surveyed," TerraChoice Environmental Marketing Inc., press release, November 19, 2007, <www.terrachoice.com/files/EcologoReleaseUS.pdf> (September 2008).

46. Ibid.

47. Ellison.

48. Ibid.

49. "Actions for Living Green."

50. Associated Press, "New Exhibit has Visitors Thinking Green."

51. "Fifty Ways to go Green," WNBC, 2007, <www.wnbc.com/print/12907475/detail.html> (February 2008).

52. "Did You Know?"

Chapter 4. Street Smart

1. Author's Interview with Jennifer Kirk, August 2008.

2. Ibid.

3. Ibid.

4. Lydia Saad, "Half of Americans Driving Less to Save Gas," Gallup News Service, April 10, 2006, <www.gallup.com/poll/22303/Half-Americans-Driving-Less-Save-Gas.aspx> (September 2008).

5. "Ratings Highlights," Greener Cars, n.d., <www.greenercars.org/highlights.htm> (April 5, 2010).

6. Saad.

7. Jesse Ellison, "Save the Planet, Lose the Guilt," Newsweek, July 7–14, 2008, <www.newsweek.com/143701> (September 2008).

8. Ibid.

9. Bryan Walsh, "The Chevy Volt: GM's Huge Bet on the Electric Car," TIME, September 16, 2008, pp. 55–56.

10. Ellison.

11. Walsh.

12. "Chevrolet Volt Leads General Motors Into Its Second Century," General Motors, press release, September 16, 2008, <http://gm.com/servlet/GatewayServlet?target=http://image.emerald.gm.com/gmnews/viewpressreldetail.do?domain=2&docid=48589> (October 2008).

13. Walsh.

14. Ibid.

15. Jim Ostroff, "Air Cars: A New Wind for America's Roads?" Kiplinger Washington Editors, Inc., October 30, 2008, <http://finance.yahoo.com/family-home/article/106040/Air-Cars:-A-New-Wind-for-America's-Roads> (October 2008).

16. Clifford Krauss, "Driving Less, Americans Finally React to Sting of Gas Prices, a Study Says," *New York Times,* June 20, 2008, <http://biz.yahoo.com/nytimes/080620/1194786578804.html?.v=19> (October 2008).

17. Ibid.

18. "What You Can Do About Car Emissions," National Safety Council, February 27, 2008, <http://www2.nsc.org/ehc/mobile/mse_fs.htm> (April 2010).

19. Frank Newport, "Americans Convinced Rise in Gas Prices is Permanent," Gallup, May 9, 2008, <http://www.gallup.com/poll/107170/Americans-Convinced-Rise-Gas-Prices-Permanent.aspx> (December 2009)

20. Krauss.

21. Michael Grunwald, "The Clean Energy Scam," *TIME,* April 7, 2008, pp. 40–45.

22. Ibid.

23. Ibid.

24. Ibid.

25. Ibid.

26. Associated Press, "Bush Signs Bill Requiring 35 mpg Cars by 2020," *Columbia Tribune,* December 9, 2007, <www.columbiatribune.com/2007/dec/20071219News022.asp> (October 2008).

27. Associated Press, "Bush Signs Bill Boosting Car Fuel Efficiency," MSNBC, December 19, 2007, <www.msnbc.msn.com/id/22326795> (October 2008).

28. Ken Thomas, "Auto Industry Shows Off Fuel Efficiency," North American International Auto Show, January 14, 2008, <www.naias.com> (February 2008).

29. Associated Press, "Bush Signs Bill Boosting Car Fuel Efficiency."

30. Trystan L. Bass, "Drive-thrus are a waste," Yahoo! Green, January 18, 2008, <http://green.yahoo.com/blog/greenpicks/101/drive-thrus-are-a-waste.html> (October 2008).

31. "What You Can Do About Car Emissions."

32. Ibid.

33. Ibid.

34. Trystan L. Bass, "Saving Gas Isn't Just For Tree-Huggers Anymore," Yahoo! Green, July 22, 2008, <http://green.yahoo.com/blog/greenpicks/191/saving-gas-isn-t-just-for-tree-huggers-anymore.html> (October 2008).

35. Joan Shim, "Six Ways to Stretch a Tank of Gas," Yahoo! Green, February 26, 2008, <http://green.yahoo.com/blog/forecastearth/42/six-ways-to-stretch-a-tank-of-gas.html> (October 2008).

Chapter 5. **Knowledge Is Power**

1. Author's Interview with Christy Radanof, September 2008.

2. Ibid.

3. Ibid.

4. Ibid.

5. Ibid.

6. Ibid.

7. "Healthy Schools: School Buses," Environmental Association for Great Lakes Education, n.d., <www.eagle-ecosource.org/buses.html> (October 2008).

8. "What Parents Need to Know about Diesel School Buses," Natural Resources Defense Council, March 17, 2001, <www.nrdc.org/air/transportation/qbus.asp> (April 5, 2010).

9. "Idling Reduction," Adopt-A-School Bus Program, n.d., <www.cleanschoolbus. net/EmissionReductionToolkit.htm> (April 5, 2010).

10. Dan Simon, "Reducing School Bus Emissions," Greenworks TV, September 10, 2002, <www.greenworks.tv/radio/todaystory/20020910.htm> (February 2008).

11. "Myths About Energy in Schools: Energy Smart Schools," Energy Smart Schools Brochure: U.S. Department of Energy, February 2002, <www.nrel.gov/docs/ fy02osti/31607.pdf> (April 5, 2010).

12. "Energy Savings Tips for Schools," Alliance to Save Energy, n.d., <www.ase.org/ content/article/detail/625> (April 5, 2010).

13. "School Operations and Maintenance: Best Practices for Controlling Energy Costs," Alliance to Save Energy, August 2004, <www.ase.org/uploaded_files/ greenschools/School%20Energy%20Guidebook_9-04.pdf> (April 5, 2010).

14. "Going Green 101: Resources for Your Family," *The Oprah Winfrey Show,* April 20, 2007, <www.oprah.com/article/oprahshow/tows_past_20070420_b> (April 5, 2010).

15. "Fun Facts about Saving Energy," Alliant Energy, n.d., <www.alliantenergykids. com/stellent2/groups/public/documents/pub/phk_ee_se_001499.hcsp#P-4_0> (February 2008).

16. "Myths About Energy in Schools: Energy Smart Schools."

17. Ibid.

18. "Commercial Buildings: Schools," Building Technologies Program, U.S. Department of Energy, September 25, 2009, <www1.eere.energy.gov/buildings/ commercial/schools.html> (April 5, 2010).

19. "Myths About Energy in Schools: Energy Smart Schools."

20. "Fun Facts about Saving Energy."

21. "School Operations and Maintenance: Best Practices for Controlling Energy Costs."

22. "Green Power Solar Schools," Electric Cooperatives of South Carolina, n.d., <www.ecsc.org/index.php?option=com_content&task=view&id=178&Itemid= 295> (April 5, 2010).

Chapter 6. Sustainable Communities

1. Ben Feller, "Bush Hails Recovery of Tornado-leveled Kansas Town," ABC News, 2008, <http://abcnews.go.com/Politics/Weather/wireStory?id=4782764> (October 2008).

2. "Devastated Kansas Town Goes Green," ABC News, August 15, 2007, <http://abc-news.go.com/GMA/story?id=3481990&page=1> (October 2008).

3. Feller.

4. Associated Press, "After Twister, Greener Greensburg Rises," MSNBC, May 2, 2008, <www.msnbc.msn.com/id/24416341/wid/18298287/> (October 2008).

5. Bryan Walsh, "Postcard: Greensburg. Building back from the rubble," *TIME,* March 17, 2008, p. 8.

6. "Devastated Kansas Town Goes Green."

7. "Recovery Plan," Official Web Site of Greensburg, Kansas, 2007, <http://www.greensburgks.org/recovery-planning/long-term-community-recovery-plan/> (October 2008).

8. Associated Press, "After Twister, Greener Greensburg Rises."

9. Brett Zongker, "Greensburg featured in national museum," Kansas Rural Water Association, November 3, 2008, <http://www.krwa.net/newsDB/MainAnnounce2.asp?key=448> (February 2010).

10. Ibid.

11. Associated Press, "After Twister, Greener Greensburg Rises."

12. "Devastated Kansas Town Goes Green."

13. "Issues: Smart Growth," Natural Resources Defense Council, n.d., <http://www.nrdc.org/smartgrowth/default.asp> (April 5, 2010).

14. "Sustainable America," Civic Renewal Movement: CPN, February 1996, <www.cpn.org/topics/environment/sustainable.html> (April 5, 2010).

15. "Benefits for the Environment," Green Communities, n.d., <www.greencommuni-tiesonline.org/green/benefits/environment.asp> (April 5, 2010).

16. "The ENERGY STAR for Buildings and Manufacturing Plants," Energy Star, n.d., <http://www.energystar.gov/index.cfm?c=business.bus_bldgs> (http://www.energystar.gov/index.cfm?c=business.bus_bldgs).

17. "Sustainable Community Examples," Rand Corporation, April 1997, <www.rand.org/pubs/monograph_reports/MR855/mr855.ch5.html> (April 5, 2010).

18. "Vermont City Nabs Eco-Friendly Honor: Country Home Magazine Survey Cites 'Best Green Places' To Live in America," CBS News, March 8, 2007, <www.cbsnews.com/stories/2007/03/08/tech/main2547648.shtml?source=search_story> (November 2008).

19. Ibid.

20. Ibid.

21. Caryn Rousseau, "Chicago Outlines Plan to Slash Greenhouse Gases," September 19, 2008, <www.enn.com/pollution/article/38231> (November 2008).

22. "Five Ways to Go Green From Al Gore," CNN.com, August 24, 2007, <www.cnn.com/2007/LIVING/wayoflife/08/23/o.green.gore> (April 5, 2010).

23. Ibid.

24. "How to Plant a Tree," American Forests, n.d., <www.americanforests.org/plant-trees/howto.php> (April 5, 2010).

25. "Actions for Living Green," Living Green, April 2007, <www.livinggreen.org/ac-tions.cfm> (April 5, 2010).

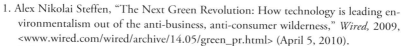

Chapter 7. **Red, White, and Green! Green Power and the Future of America**

1. Alex Nikolai Steffen, "The Next Green Revolution: How technology is leading environmentalism out of the anti-business, anti-consumer wilderness," *Wired,* 2009, <www.wired.com/wired/archive/14.05/green_pr.html> (April 5, 2010).

2. Ibid.

3. "The Greening of America's Campuses," *Business Week,* April 9, 2007, <http://www.businessweek.com/magazine/content/07_15/b4029071.htm> (November 2008).

4. "Our Story," Step It Up 2007, <http://stepitup2007.org/article.php?list=type &type=48> (April 5, 2010).

5. "The Greening of America's Campuses."

6. "World Can Halt Fossil Fuel Use by 2090," ABC News, October 27, 2008, <http://abcnews.go.com/Technology/story?id=6119855&page=1> (April 5, 2010).

7. Ibid.

8. Ibid.

9. Jesse Ellison, "Save the Planet, Lose the Guilt," *Newsweek,* July 7–14, 2008, <www.newsweek.com/143701> (September 2008).

10. "Solutions," Cooler Heads Coalition, n.d., <www.globalwarming.org/solutions> (November 2008).

11. Ibid.

12. "Energy & Environment," White House Web Site, 2009, <www.whitehouse.gov/issues/energy-and-environment> (December 2009).

13. H. Josef Hebert, "Report: 4.2 Million New 'Green' Jobs Possible," ABC News, 2008, <http://abcnews.go.com/US/wireStory?id=5930900> (November 2008).

14. Ibid.

15. RP News Wires, "Americans Feel Energy Woes Won't Be Solved in Their Lifetime," n.d., <www.reliableplant.com/articl.asp?pagetitle=Americans+feel+energy+woes+won't+be+solved+in+their+lifetime&articleid+1670> (November 2008).

16. Ibid.

17. Author's Interview with Craig Pickerill, September 2008.

Glossary

biofuel—A type of fuel that is made from plant materials.

biomass—Natural material such as wood, paper, or waste that is used to make energy.

cap-and-trade system—A system that puts a cap, or ceiling, on how much carbon a company can produce; companies that produce less carbon can sell or trade to those who are over their limit.

carbon dioxide—A gas that is released into the atmosphere by both natural processes and human activities including driving, construction, daily living, and more; plants and trees store carbon dioxide.

carbon footprint—A calculation of how much carbon dioxide is released into the environment by an individual, a group, a building, or another entity.

carbon offset—A financial tool used to offset the size of a person's carbon footprint; an example would be paying to have trees planted as a way to offset the carbon a person puts into the air when driving a car.

climate change—A significant change in temperature as well as changes in wind and precipitation.

compressed air vehicle (CAV)—A car that works by compressing air instead of gasoline to move engine pistons.

daylighting—A process in which light is captured from the rooftop of a building and redirected to interior spaces.

deforestation—The process of cutting down and sometimes burning trees; increases carbon dioxide in the atmosphere.

emission—A substance discharged into the air.

Energy Star—A symbol used for energy efficiency.

energy vampires—Another name for electronic devices that use power even when they are turned off.

ethanol—A fuel made from corn kernels.

fossil fuel—Fuel that is made from the fossilized remains of things that died years ago; oil and coal are examples of fossil fuels.

global warming—An increase in the average temperature of the earth's atmosphere.

greenhouse effect—A natural process in which some energy from the sun is absorbed and turned into heat to warm the earth. Global warming increases the greenhouse effect by making it harder for heat to be reflected back to the sun.

greenhouse gas—A gas found in the atmosphere, like carbon dioxide, that contributes to the greenhouse effect.

green power—Energy that provides the highest environmental benefit; also see renewable energy.

greenwashing—A term used to describe a company's methods to try to appear "greener" than it really is.

hybrid—A type of vehicle that has a battery-powered motor that works with a gasoline-powered engine to reduce the amount of fuel that is needed.

Kyoto Protocol—An agreement reached by the United Nations Conference on Climate Change in Kyoto, Japan, in 1997.

nonrenewable energy—Resources that cannot be replaced, like oil and coal.

off-gas—Chemicals released into the air when a product is new.

renewable energy—Resources that can be replaced; unlike coal or oil, it will not run out; also called "clean energy" or "green power."

sustainable—Capable of being continued with minimal long-term effect on the environment.

Zero Energy home—A home containing energy-efficient construction and appliances and renewable energy systems.

For More Information

Alliance to Save Energy
1850 M Street N.W.
Suite 600
Washington, D.C. 20036
202-857-0666

Environmental Defense Fund
257 Park Avenue South
New York, NY 10010
212-505-2100

Green Building Council
2101 L Street N.W.
Suite 500
Washington, D.C. 20037
1-800-795-1747

Natural Resources Defense Council
40 West 20th Street
New York, NY 10011
212-727-2700

Sustainable Buildings Industry Council
1112 16th Street N.W.
Suite 240
Washington, D.C. 20036
202-628-7400

United States Department of Energy
United States Environmental Protection Agency
Ariel Rios Building
1200 Pennsylvania Avenue N.W.
Washington, D.C. 20460
202-272-0167

Further Reading

Books

Foland, Andrew Dean. *Energy*. New York: Chelsea House Publishers, 2007.

Fridell, Ron. *Earth-Friendly Energy*. Minneapolis: Lerner Publications, 2008.

Rau, Dana Meachen. *Alternative Energy Beyond Fossil Fuels*. Minneapolis, : Compass Point Books, 2010.

Reilly, Kathleen M. Energy: *Why We Need Power and How We Get It*. White River Junction, Vt.: Nomad Press, 2009.

Sherman, Jill. *Oil and Energy Alternatives*. Edina, Minn.: ABDO Publishing, 2009.

Internet Addresses

Alliance to Save Energy
<http://ase.org/section/_audience/consumers/kids>

Alliant Energy Kids
<http://www.alliantenergykids.com/index.htm>

Kids Saving Energy
<http://www.eere.energy.gov/kids/>

Index